CURVED-FOLDING
ORIGAMI DESIGN

CURVED-FOLDING
ORIGAMI DESIGN

JUN MITANI

CRC Press
Taylor & Francis Group
Boca Raton London New York

CRC Press is an imprint of the
Taylor & Francis Group, an **informa** business

AN A K PETERS BOOK

CRC Press
Taylor & Francis Group
6000 Broken Sound Parkway NW, Suite 300
Boca Raton, FL 33487-2742

© 2019 by Taylor & Francis Group, LLC
CRC Press is an imprint of Taylor & Francis Group, an Informa business

No claim to original U.S. Government works

Printed on acid-free paper

International Standard Book Number-13: 978-0-367-18027-0 (Hardback)
978-0-367-18025-6 (Paperback)

Library of Congress Cataloging-in-Publication Data

Names: Mitani, Jun, 1975- author.
Title: Curved-folding origami design / Jun Mitani
Other titles: Curved folding origami design
Description: Boca Raton, Florida : CRC Press, c2019. | Includes
bibliographical references and index.
Identifiers: LCCN 2018060524| ISBN 9780367180256 (pbk. : alk. paper) | ISBN 9780367180270 (hardback : alk. paper) |
ISBN 9780429059179 (ebook : alk. paper)
Subjects: | LCSH: Origami. | Folds (Form) | Paper work |
Classification: LCC TT872.5 .M5644 2019 | DDC 736/.982--dc23
LC record available at https://lccn.loc.gov/2018060524

Visit the Taylor & Francis Web site at
http://www.taylorandfrancis.com

and the CRC Press Web site at
http://www.crcpress.com

Contents

Preface

I think most of you enjoyed origami when you were kids. The most basic origami is to fold a sheet flat along a straight line, corner to corner or side to side. If you were to "fold this paper as you like," you would start with folding it flat. It is widely considered that paper should be tightly folded. However, paper is bendable. The paper can also be folded along a curve by adding a fold on its bended, curved surface. This "curved folding" has not drawn much attention. However, the surface of the paper created from curved folding produces geometric, elegant shading. I have created a number of geometric solid objects through computer-calculated designing of shapes obtained from folding curves. Those techniques are organized in the book titled *3D Origami Art* (CRC Press, 2016). Fortunately, that book inspired many people to know the world of curved-fold origami. At the same time, it made me recognize again that computer-based designing is a high hurdle and not an easy method. Do we really need a computer to make shapes by folding paper along curves? I am a computer science researcher, but I gave it a try with computer-free curved-fold origami. Paper can be folded along curves relatively freely, but not *that* freely. Most randomly drawn curves do not fold well. To obtain beautifully designed modelings, we at least need knowledge about how fold lines are laid out. This book presents shapes that are formed by curve-folding paper, together with fold line patterns. Its contents are a record of fun experiments from my own real experience with sheets of paper. You will be able to create a variety of shapes by combining the fold line patterns in this book. I hope to share the beauty of curved surfaces created by paper with as many people as possible.

Jun Mitani

Prologue: Folding a Curve Means a Lot

P.1 Shapes from Bending Paper

First, let's be clear about the difference between "bending" and "folding" a sheet of paper. Bending is to change the entire paper into a smooth shape without creating edges, as in Figure P.1.

This produces a smoothly curved surface. However, the bent shape quickly opens back flat if it is not held or taped due to paper's ability to return to its original state. The end result is the shape balanced between this property of the paper and the fixed point. A somewhat thick sheet produces a beautifully curved, tense surface with no deflection.

Paper is inelastic and does not deform freely like a rubber sheet. There are only three known curved surfaces that can be made from bending a sheet of paper—cone, cylinder, and tangent surfaces. These are called "developable surfaces". A developable surface is a shape consisting of lines. Each shape in Figure P.2 is also an assembly of lines, though they are not visible. So, a ruler tightly fits on its surface. Figure P.3 illustrates the consisting lines of the three types of developable surfaces. By actually bending a sheet and closely observing its surface, you can roughly estimate how the lines are arranged. And, thinking the direction of these lines are very important in creating a curved surface out of a sheet.

There are only three types of developable surfaces. However, you can make shapes with different types of developable surfaces, as seen in Figure P.4, by smoothly connecting a curved surface with another curved one or a curved surface with a plane. The left is estimated to be a shape that combines a cylinder

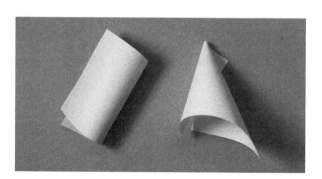

Figure P.1   Bent paper.

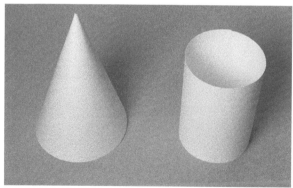

Figure P.2   Paper bent into a cone and a cylinder.

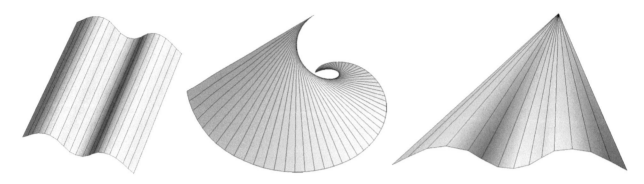

Figure P.3   Curved surface made from paper is an assembly of lines.

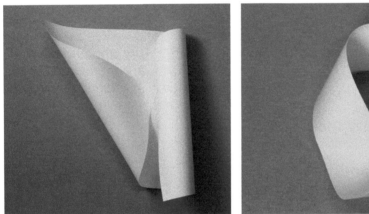

Figure P.4   Bending a sheet makes various shapes (partly taped).

and plane while the right a complex combination of cones. Just bending a sheet makes a variety of shapes.

P.2 Shapes from Folding Paper

Folding Lines

Folding produces an edge on the paper. For instance, folding the paper in half and pressing down on the folded side produces a straight line. Folding paper flat is something so common in our everyday life, as seen in the left photo of Figure P.5. Unfolding the folded paper leaves the folded line as an acute edge,

because the paper retains the folded state. The final shape is the balance between forces to open back and to stay folded, like the right photo in Figure P.5. The more firmly folded, the more acute the ridge is. Repeatedly, straight folding is to fold the paper along a line and produce a sharp straight ridge.

Folding Curves

Paper can be folded along curves. The result is a curve fold, as in Figure P.6. Press-folding flat gives just straight folds. To make a curve fold, you need to fold the paper in the air as you bend the entire piece. This is to curve-fold the paper. Curved smooth surfaces are formed across the crease after a curved fold.

Figure P.5   Paper folded along a line.

Figure P.6   Paper folded along a curve.

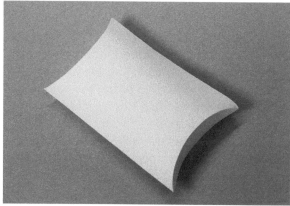

Figure P.7   Pillow-shaped package.

It can be said that two curved surfaces are connected without a gap at the crease. Like in straight folding, the paper retains the folded state and remains in a curved shape even after you let go of it. The paper settles in a shape where the forces that retain the crease and return to the flat state are balanced. A balanced state creates a beautiful smooth shape. Partially pinching or taping the folded paper also creates another stable state. I think everyone has seen a pillow-shaped package, as in Figure P.7. This is a good example of making effective use of curve-folding.

P.3 Crease Pattern

Unfolding the creased paper back to the flat state leaves folding traces on the paper. The graphic showing these traces is called a "crease pattern". Folding traces in the crease pattern are called "fold lines". There are two-fold line types—the "mountain" fold and the "valley" fold. A mountain fold literally looks convex in the front while a valley fold looks concave. The mountains and valleys are reversed on the back of the folded paper. In crease patterns, mountains are shown by solid lines and valleys by dashed lines for visibility.

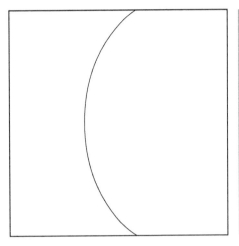
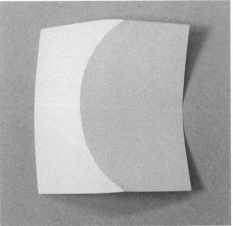

Figure P.8   Crease pattern containing a mountain fold curve and resulting shape.

Contours of the paper are shown by thick sold lines. Shapes photographed in this book can be reproduced by tracing the crease pattern onto a sheet and folding along the fold lines.

Note that the fold line has no information about the degree of folding angle. The crease pattern does not tell whether a line should be folded sharply or lightly. If a sheet is to be folded flat, it means the fold angle is 180 degrees. So, it is enough to show whether it is mountain or valley. On the other hand, when creating a three-dimensional shape, the fold angles vary and the outcome depends on the angles. However, because it is very difficult to tell the degree of folding angle in the crease pattern, this book's crease patterns provide only mountains and valleys like other origami books do (Figure P.8). Please see the finished photo for the folding angles. For example, the pillowcase crease pattern is shown in Figure P.9. All fold lines are mountains. Use of the crease pattern and finished photo will work to some degree to reproduce the shape.

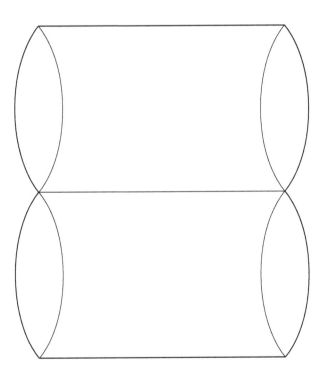

Figure P.9   Pillowcase crease pattern.

P.4 How Line Elements Are Arranged on a Curved Surface

A curved surface made from bending the paper is an assembly of lines as explained in Section P.1. Lines that consist of a curved surface are called "line elements" or "rulings". It is important to imagine how these line elements will be arranged on the shape you are making. The line elements change direction at the curved fold in alignment without crossing each other, as in Figure P.10. Generally, line elements are arranged at almost right angles to the folds.

Conversely, curved surfaces cannot be made that have intersecting line elements. So, we have to be very careful not to cause line elements to cross when making a crease pattern or folding the paper. Still, it is not easy to predict how the line elements are aligned. This is one of the reasons for difficulty in designing shapes containing curved folds.

However, the knowledge that a shape made from bending the paper has line elements will help you notice earlier whether the shape is possible or impossible, eliminating a lame try. "For those who want more theory—2" on page 28 explains the fold line shapes and the resulting shape's projections and recesses for further information.

P.5 Are Any Curves Foldable?

As an experiment, draw a random curve on paper and fold along it. You'll easily get a shape with a curved surface. Now the question is, do any curves work well on paper? Simple curves like those in Figure P.8 are neatly foldable, but squiggles such as those in Figure P.11 are not. Unsuccessful cases contain multiple curves, multiple intersecting curves, and lines with sharp edges. Give it a try. You'll find that the curves, as listed below, cannot fold to a shape with a smoothly curved surface.

- Too squiggly
- Unfinished or looped
- Multiple intersecting curves
- A set of simple lines that looks good, but in bad arrangement of mountains and valleys

In many cases, a curve drawn without some prior knowledge cannot be folded to a solid with a curved surface. It's much more difficult when multiple curves need to be combined.

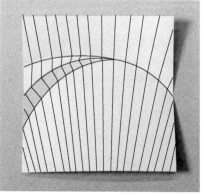

Figure P.10   A curved-fold shape with folds and line elements drawn on it.

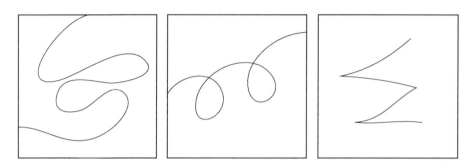

Figure P.11   Fold lines not neatly foldable.

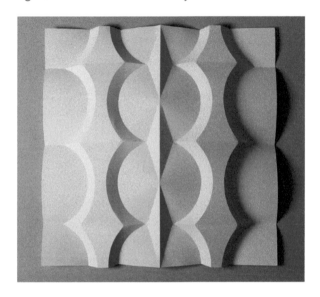

Figure P.12   Work combined with curved folds.

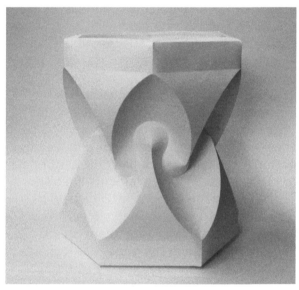

Figure P.13   The Huffman Tower. (Reproduced by the author.)

The symmetrical convex-concave folds, like those in Figure P.12, may have a certain rule. This book presents the curve layouts for producing beautiful shapes organized in chapters. By combining them, you will create various curved-fold origami shapes different from traditional ones.

P.6 Past Efforts on Curved-Fold Origami

This book introduces how to design curved-fold origami without a numerical formula. However, mathematics about curved paper

surfaces has attracted many researchers. David Huffman (1925–1999), the computer scientist famous for inventing the Huffman code, greatly contributed to creating many curved surface origami works. Figure P.13 is one of his masterpieces, "The Huffman Tower" (reproduced by the author). Huffman's design approaches have been analysed by computer scientist Erik Demaine. He and his father, Martin Demaine, have created a lot of curved-crease sculptures which use the closed curved-fold techniques explained in Section 7.4. Ron Resch (1939–2009), computer scientist, was an early spotter of curved paper surfaces and computer-aided shape designing. Architect Roy T. Iwaki (1935–2010) created vivid three-dimensional origami masks of animals folded from a sheet of paper. His techniques are collected in his book titled *The Mask Unfolds* (2000). It presents specific know-how for creating desired shapes, including how curves should be laid out.

Nowadays, you can see original origami creations in Flicker* and Instagram† posted by origami lovers from around the world. Quite a few of them are curved-fold origami works. Searching by "curved fold" will find many interesting works. Curved-fold origami has also become an active area of academic research in geometry, design, and simulation. It is one of the hottest topics at the quadrennial International Meeting on Origami in Science. Origami containing curved folds is expected to become more popular in the near future.

This book's contents are based on the materials and literature as well as the know-how gained through my experience using my hands in an organized and understandable manner in order to make you feel easy to try curved-fold origami creation.

---

* https://www.flickr.com/
† https://www.instagram.com/

Tips to Fold Neatly

I would like to end this chapter with some tips to help you make the origami creations introduced in this book.

Paper Used

Using an appropriate type of paper makes a big difference. Choose a sheet that is somewhat thick and tends to retain its folded state. Texture greatly affects the look of the outcome. I prefer TANTO paper which is available in Japan. A thickness of around 100 kg ream weight is good. All of my works photographed in this book use 100 kg TANTO paper.

Tools Used

Origami is tool-free. However, some tools will do wonders in folding curves neatly. Using all or part of the following tools helps you neatly make the examples and works presented in this book.

- Straight ruler, curve ruler: Used to draw a smooth curve or line and make a pre-crease.
- Stylus: Used to make a pre-crease. A sharp pointed tool will do, such as a ballpoint pen.
- Scissors, box-cutter: Used to cut the paper to the necessary size.
- Cutting mat: Placed under the paper when cutting with a box-cutter or when applying a stylus.
- Glue, tape: Used to partly fix and stabilize the folded paper.
- Cutting plotter: A PC-controlled paper cutting machine. Used also for pre-creasing when the crease pattern is available in digital file formats like Adobe Illustrator.

Get the Crease Pattern

This book provides the basics for curve-folding a shape, followed by the sample work created through combinations of techniques. Each basic shape comes with a simple crease pattern and the photograph of the folded result. Exact reproduction is not necessary. Use a curve ruler and try to draw curves similar to those provided in this book.

You may want to use the crease pattern of each sample work photocopied on a thick sheet of paper or print out the data downloaded from the Internet. Visit my webpage at http://mitani. cs.tsukuba.ac.jp/book/curved_folding/ to find the crease patterns of all of the sample works in this book.

Pre-crease the Lines

Unlike those in general origami books, the pre-creased origami works in this book are basically folded along curves. It may not go smoothly in the beginning. Just keep at it and try many times. Eventually, you will achieve a satisfactory outcome. To finish your work beautifully, it is important to pre-crease the fold lines firmly. Without this, it is impossible to fold a sheet along the intended lines. Sharp pre-creases make the workpiece neat at each step for the succeeding folding process. It can be said that pre-creasing counts the most in the process.

A cutting plotter machine can automatically and neatly pre-crease the fold lines. This machine makes pre-creases by lightly tracing the surface with the cutter blade with a controlled force, or it may also be set up to finely perforate the fold lines. One example of an affordable small cutting machine, as of this writing, is the Graphtec Silhouette CAMEO, which is capable of handling up to A4-size sheets (or letter size). For a larger work, a cutting plotter supporting up to A3 size is a good choice (e.g., Craft ROBO-PRO).

If you do not have a plotter, use a sharp pointed tool, such as a stylus, to press firmly along the fold lines. A ruler is good enough to pre-crease straight lines, but not so for curved lines. A curve ruler is a nice idea. The best way to use it is to cut out the curved portion of the crease pattern from a thick sheet and use it as the special ruler, because curves on a crease pattern are often in the same shape.

Folding Process

Once you make sharp pre-creases on the sheet, you have done more than half toward a beautiful finish. In the succeeding folding process, it is difficult to fold curves until you get used to it. Folding on a desktop gives a straight fold. To fold a curve, hold the sheet up with both hands and apply the intended fold as you warp the sheet. Fold the lines as sharply as possible so the sheet memorizes the fold state. Avoid making unnecessary folding traces. Carefully fold the necessary portions only. A crease pattern of the wrong size (too small or too large) cannot be comfortably handled. A crease pattern in a size that is controllable with both hands may be a good one to use.

Finishing

Gluing may be contrary to the origami rules, but sometimes is necessary to stabilize the final shape. It is difficult for paper to be folded or kept exactly at the intended angle. Fixing with wood glue or double-sided tape stabilizes the entire piece.

Photographing

Take a photo of your finished work. Though it depends on your preference, a work made from white paper photographs beautifully in three-dimensional black and white shading. Try with various light positions. Looking behind the work may change the entire impression.

These are the tips for making a beautiful work. Again, the origami creations in this book are quite unusual. Never give up and try many times, and you'll find yourself capable of doing wonderfully. Give yourself a hands-on experience and enjoy folding the paper.

Author

Jun Mitani received his Ph.D. in engineering from the University of Tokyo in 2004. He has been a professor at the University of Tsukuba since April 2015. His research interests center on computer graphics—in particular, geometric modeling techniques and their application to curved origami—as well as interactive design interface. *3D Origami Art* (Taylor & Francis/CRC Press) is one of his published books.

Chapter 1

Fold a Curve

For the first step, fold a single curve and see the shape you get. If this is the first time ever for you folding a curved line, then I'm pretty sure it will be a whole new entertaining experience. You will be surprised at the shape variations produced by a single curve.

1.1 Folding a Simple Curve

First, let's fold a simple curved line. A simple curve runs in one direction, like a parenthesis "(". It does not turn to a different direction halfway like the letter "S". Now, let's draw a curve on the paper, as in Figure 1.1, and fold along it. Draw the curve with a hard-pointed

tool to pre-crease. A curve ruler helps draw a smooth curve, but freehand will do at first.

Figure 1.1 shows the paper contour in a thick solid line and the curve to be folded in a thin solid line. The solid line is folded as a mountain. You may want to fold it as a valley, but, as a practice for understanding crease patterns, mountain-fold it as per the figure.

Start with holding up the paper, not placing it on the desktop, and bending the entire piece. Make a crease as you bend the paper with both hands, as shown in Figure 1.2. Be careful not to press flat, otherwise the resulting folded line becomes straight. If pre-creased firmly, the line should fold beautifully. Avoid wrinkles and undesired folds. Once folded successfully, the paper becomes a shape with one side raised and another recessed across

Figure 1.1   Crease pattern with a simple curve.

Figure 1.2   Folding a curve while bending the entire piece.

the fold line. Once you made the shape, as in Figure 1.3, hold the workpiece against a light at various angles. The curved surface produces beautiful gradual shading varying with the angle of illumination.

Next, pinch the fold line at both ends and twist the workpiece freely, as in Figure 1.4. Paper is re-shapable, even after curve-folding. Avoid excessive force, otherwise wrinkles form. Twisting the entire piece spirals the fold

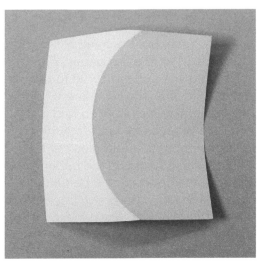
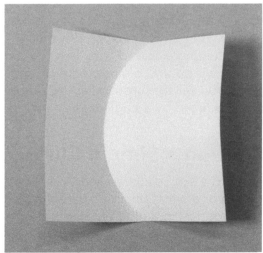

Figure 1.3   Figure 1.1 crease pattern folded (right: backside).

Figure 1.4   Workpiece entirely twisted into another shape.

line, producing an unexpected shape. The entire shape is changeable while remaining smooth. When you release your hands, the workpiece settles at the balance between the paper's forces to return to its original state and those of the folded line. You can see the workpiece's fold line settling on a plane.

By folding the line more sharply and bending the entire workpiece, the resulting shape becomes more three-dimensional. Partly fix the workpiece, instead of releasing your hands, to allow it to stabilise at the state seen in Figure 1.5 which shows cylinders with a larger bend and glued edges. The folded back portion opens outward. Figure 1.6 shows cone-like shapes with one end of the cylinder tapered. The right photo looks like a witch hat. It's just folding a curve. But it makes all

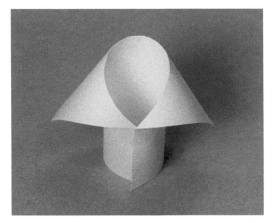 

Figure 1.5   Bent largely to make a cylinder.

 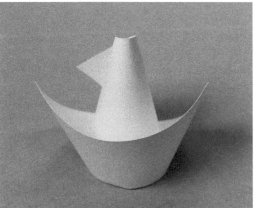

Figure 1.6   Bent largely to make a cone.

  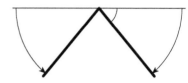

Figure 1.7 Rightmost piece folded at the largest angle.

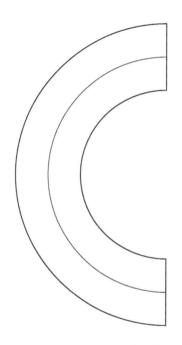 

Figure 1.8 Shape varies by degree of fold. Larger fold angle bends more, resulting in a smaller shape.

kinds of lovely shapes depending on how the whole piece is bended at whichever degree.

The greater the fold angle is, as arrowed in Figure 1.7, the larger the fold bends, distorting it into a smaller shape. The crease pattern of Figure 1.8 with a semicircle fold line will make different shapes due to the fold angle. The fold angle becomes larger from left to right on the photo. The rightmost piece is sharply bent into a smaller size. The same crease pattern produces different finishes depending on the way the paper is bent and the folding angle. Try with different combinations.

1.2 Folding a Winding Curve

Now, let's fold a winding line that changes the direction halfway (like an "S"), as in Figure 1.9. When testing, the curve may not necessarily be exactly identical to the one illustrated. Draw a smooth unjagged curve. Give it a try.

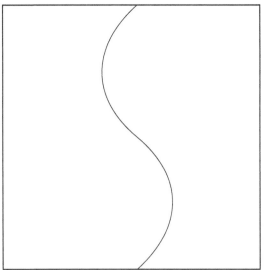

Figure 1.10   Folding Figure 1.9 crease pattern.

Figure 1.9   Crease pattern containing a winding curve.

You'll find it not easy to fold neatly because you have to bend the paper wavily. Avoid forcing the fold. Instead, twist the entire paper to allow it to fold itself. Hold the paper with both hands, as in Figure 1.10. Apply proper force as you carefully watch the paper bending. The folded result should look like Figure 1.11. The paper has two wavy curved surfaces across the fold. One side is raised, and the opposite side crossing the fold is recessed.

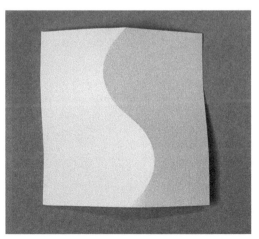

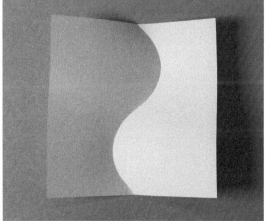

Figure 1.11   Figure 1.9 crease pattern folded (right: backside).

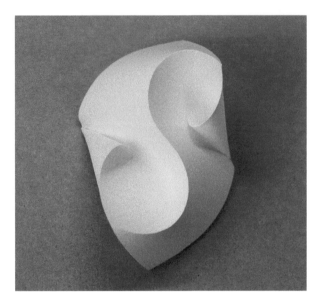

Figure 1.12 Figure 1.9 crease pattern folded, with recessed portions curled into cone.

Try a larger fold angle. You will get a more three-dimensional shape by partly fixing the folded paper. For instance, curling at the recessed portion produces an interesting two-cone shape, as in Figure 1.12. Next, fold a wavy line consisting of continuous S letters, like Figure 1.13 crease pattern. The more bends a fold line has, the more difficult it becomes to fold. Make the shape nice and easy. Hold the folded piece for a while to allow the paper to settle in the bend state. Now you get a shape with many projections and recesses. The shape produces a beautiful shading depending on the way it is illuminated. Try with many other curves. Some curves are folded well and others are not. Think about this difference.

1.3 Folding a Squiggle

Even curves drawn just any way are folded neatly, but too squiggly lines are not. Figure 1.14 is an attempt to fold a sharply winding curve. This can be folded a little but not beyond a certain fold angle. Whether a curve is neatly foldable depends not only on its shape but on the fold angle. As we see in Figure 1.8, a larger fold angle greatly changes the fold line from its original shape. Conversely, a small fold angle leaves the paper less changed from its original state without force distorting it. The distance of a fold line from the paper edge also affects the beauty of the fold. No matter how squiggly the fold line is, it is neatly foldable if it is near the edge of the paper. For instance, the squiggle of Figure 1.14 can be folded sharply

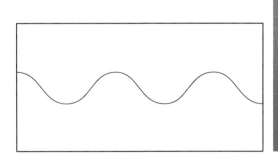

Figure 1.13   Wavy line folded.

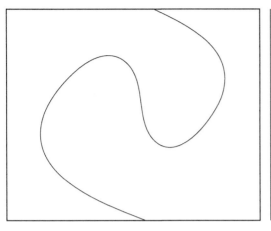
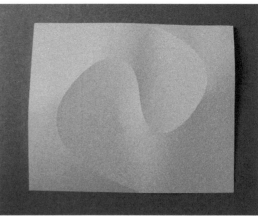

Figure 1.14   Attempt to fold a squiggle on a rectangular sheet.

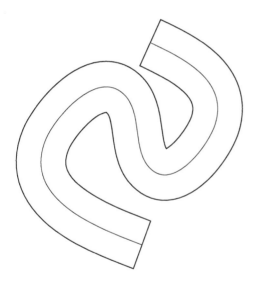

Figure 1.15   Figure 1.14 squiggle folded on a narrowed sheet.

and neatly if cut out on a narrow sheet, as in Figure 1.15.

Figure 1.16 shows a U-shaped curve folded on a narrow sheet. A U-shaped curve is foldable only when it is on a narrow sheet like this, unless otherwise add a new fold line. The larger the fold angle is, the shaper the bend becomes, causing the ends to cross, as in the photo. Figure 1.17 is an example of continuous hairpin bends, which is a combination of U-shaped curves, as shown in Figure 1.16. A sharply winding curve also becomes foldable if it is on a narrow sheet of paper. Folding like Figure 1.16 crosses the U shape's one end over another. Notice that the top end of the crease pattern comes down the side of the photo. Figure 1.18 is an example of an unsmooth, partly pointed curved line. The edge is not neatly foldable as is. Adding a valley fold line inside the corner ensures a sharp finish.

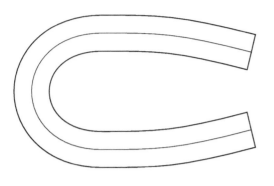 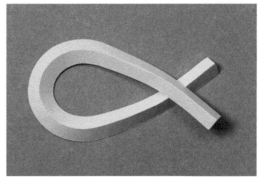

Figure 1.16   U-shaped curve folded.

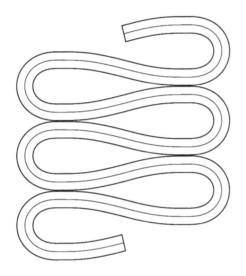 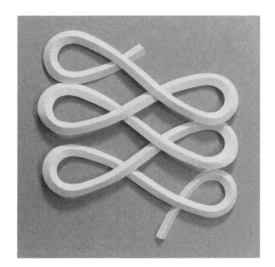

Figure 1.17   Shape from folding a snaky line.

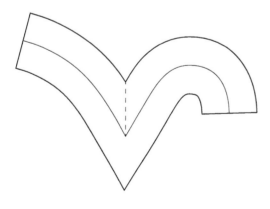 

Figure 1.18   Additional fold line inside the edge ensures the beautiful fold.

Artwork 1: Lysichiton

This example demonstrates that just folding a simple curve makes a three-dimensional work. After folding the curve shown in the figure, twist the entire piece into a shape like a lysichiton (or skunk cabbage). The inside of the curve makes the central cone of the work. To prevent the workpiece from opening, the paper is glued to fix at some portions where the edges come in contact with each other.

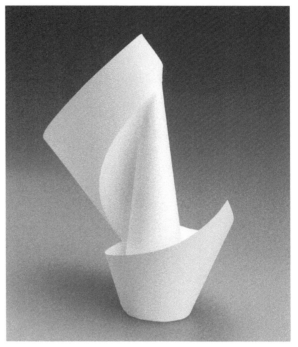

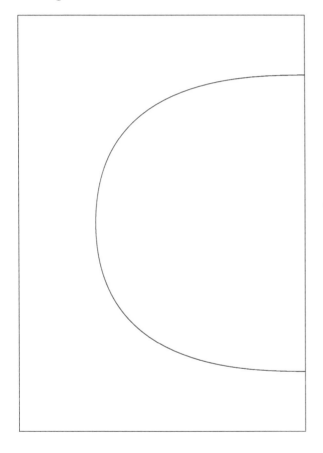

Artwork 2: Shape from Folding a Wavy Line

A wavy curve comes at the center, like the example in Figure 1.13. The work is finished with the bottom half raised and the top half tapered inwardly. It is partly fixed inside to keep the folded state. The final shape greatly changes depending on how the paper is bent on both sides of the single curve fold.

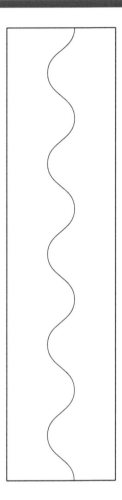

Artwork 3: Treble Clef

The degree of bend after folding depends on the fold angle, as shown in Figure 1.8. For a fold line that does not intersect on the crease pattern, the paper strip will be overlapped by folding the line at a large angle to make tight curves. Making use of this, a treble clef shape is created from the crease pattern below.

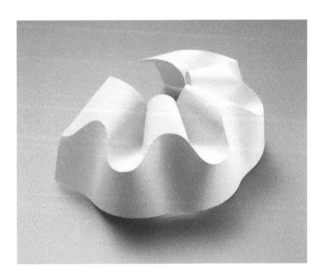

Artwork 4: Paper Braid

Two squiggles shown in Figure 1.17 are connected, then made into a long string. By adjusting the fold angle, the final shape has an almost flat central part and tight U turns. Putting the strip through the loops worked out like a knitting.

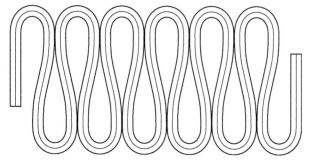

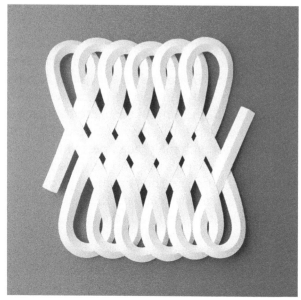

For Those Who Want More Theory–1

LINE ELEMENTS AND PAPER WIDTH

As explained on page XIII, a curved surface made from bending the paper is an assembly of line elements. These lines are invisible. If we can estimate their alignment, it is very advantageous when making new shapes. Curved surfaces cannot be made that have intersecting line elements. So, knowing of such intersections in advance will eliminate a fruitless effort with an impossible shape. This may be intuitively predicted for simple shapes, though difficult for complicated ones with multiple curves influencing each other.

As a practice, let's imagine how the line elements are aligned on the paper folded at a single curve. First, imagine lines extending from the fold to the outside of the curve. Then, imagine those lines extending toward the inside of the curve. Actually, the line elements open outward outside the curve and gather toward a point inside, as in Figure 1.19. If the fold angle is large, the lines will cross, as seen in the left diagram in Figure 1.19, becoming unfoldable. To avoid this problem, the fold angle should be smaller. Cutting off the point where lines cross is also a very easy solution, as seen in the right diagram of Figure 1.19. The "narrow paper" technique presented in this chapter is exactly for avoiding this line element crossing. Squiggles are also neatly foldable on a paper strip that is thin enough.

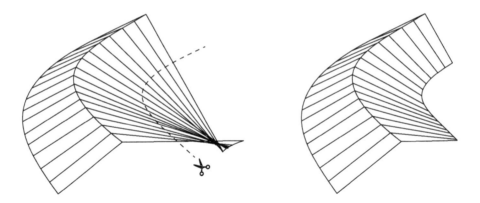

Figure 1.19   Small width can fold where it is unable to bend largely.

Chapter 2

Aligning Curves

We have worked with a single curve so far. By folding multiple curves, you will create more complex shapes. This chapter sees what shapes we can get by aligning curves in the same shape. Just "aligning" extends the range of representation.

2.1 Pleat-Folding Two Aligned Curves

Align two same-shaped curves in parallel. Figure 2.1 is an example of aligning two pieces of the simple curve shown in Figure 1.1. In the left crease pattern, the left curve is valley-folded and the right curve mountain-folded. The folds are reversed in the right crease pattern.

Give it a try. You'll find that folding both of two curves either mountains or valleys never works neatly. One curve needs to be a mountain fold and another a valley. The crease pattern folds to a terrace shape, as seen in Figure 2.2. This is called pleat fold when folding straight lines. Notice that the folded shape greatly differs depending on how the mountain and valley are laid out. The left side is recessed and the right side is projected. The mountain and valley are reversed on the back of the paper. So, one shape of Figure 2.2 becomes the same as another when flipped.

As we tried in the previous chapter, a curved surface can be rolled into a cylinder or cone shape. A crease pattern with two curves side by side can make a cylinder or cone with a step. Figures 2.3 and 2.4 fold the terrace

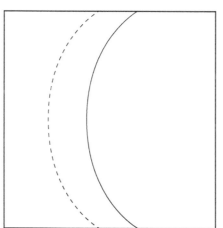 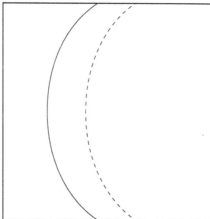

Figure 2.1   Crease pattern with two same-shaped curves aligned.

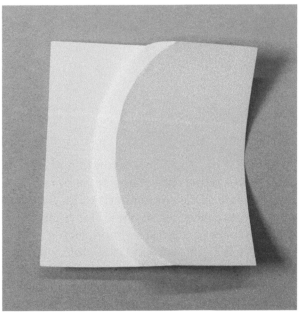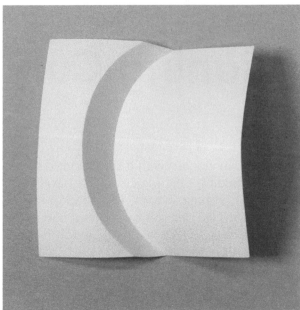

Figure 2.2   Figure 2.1 crease pattern folded.

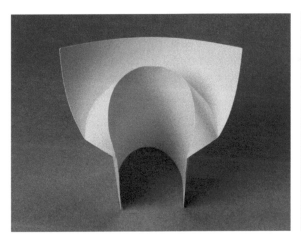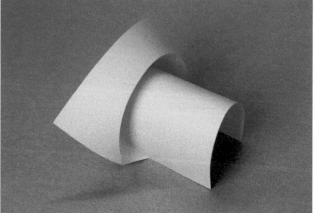

Figure 2.3   Right crease pattern in Figure 2.1 folded to a cylindrical shape.

shapes into a cylinder and a cone, respectively. Each looks quite different from different angles. Aligning just two curves in parallel will make lots of variations by changing the curve shape, paper size, curve spacing, fold angle, and degree of bend.

Next, align two pieces of a winding curve in parallel, as in Figure 2.5.

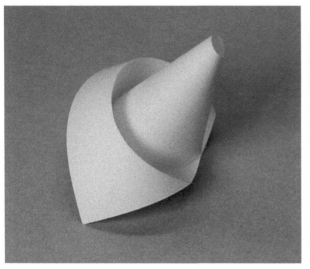
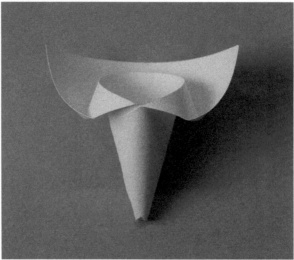

Figure 2.4   Right crease pattern in Figure 2.1 folded to a conical shape.

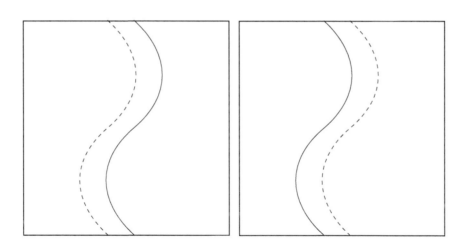

Figure 2.5   Crease pattern with two winding curves side by side.

One is mountain-folded and another valley-folded. The crease pattern with winding curves parallel aligned can also be easily folded to a beautiful shape, as seen in Figure 2.6. This method of "parallel aligning two same curves with alternating mountains and valleys" is convenient to terrace the surface.

Figure 2.6   Figure 2.5 crease pattern folded.

2.2 Aligning Many Curves

Let's align many same-shaped curves. Place the mountain and valley folds alternately. Figure 2.7 is the crease pattern where six same-shaped curves are parallel aligned, with alternating mountains and valleys. The mountain and valley assignment differs between the left and right crease patterns. The crease pattern folds to a beautiful shape, as seen in Figure 2.8, with pleat folds side by side. It presents stripe shading. Figure 2.9 is a shape from folding at a larger angle and rolling cylindrically. This brought more wave-like modeling to the shape.

There is much room for experiment. Align the curves in an unequally spaced manner, as in Figures 2.10 and 2.11. Alternate the mountain and valley folds. The distance

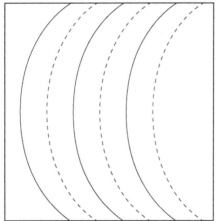 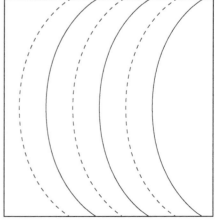

Figure 2.7   Crease pattern where multiple simple curves are aligned.

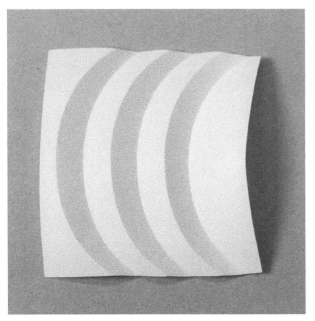

Figure 2.8   Figure 2.7 crease pattern folded.

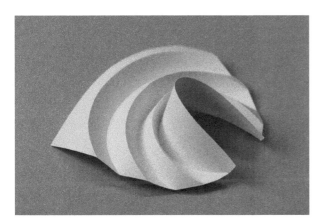

Figure 2.9   Fold lines on Figure 2.7 crease pattern are folded at a large angle.

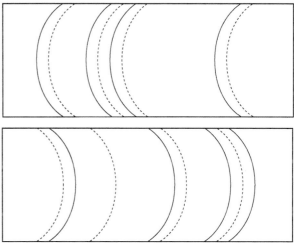

Figure 2.10   Crease pattern where simple curves are aligned in an unequally spaced manner.

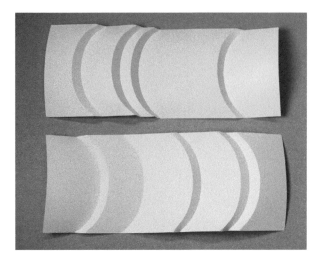

Figure 2.11   Figure 2.10 crease pattern folded.

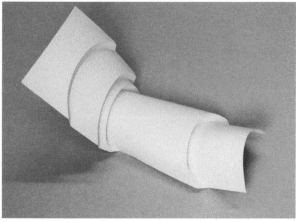

Figure 2.12   Folded at a larger angle.

between fold lines add contrast to the shading. Folding at a large angle and rolling the entire piece produces a shape like an insect leg, as shown in Figure 2.12.

Let's align many winding curves. Place the mountain and valley folds alternately, as in Figure 2.13. Careful folding makes a beautifully shaded pattern. Just reversing or turning the piece gives different impressions on the shading, as seen in Figure 2.14. Figure 2.15 is an example of winding curves aligned in an unequally spaced manner.

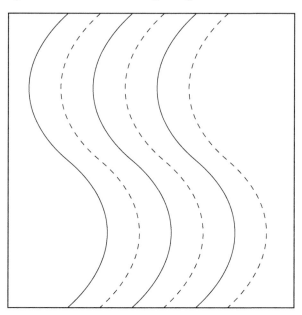

Figure 2.13   Crease pattern with winding curves side by side.

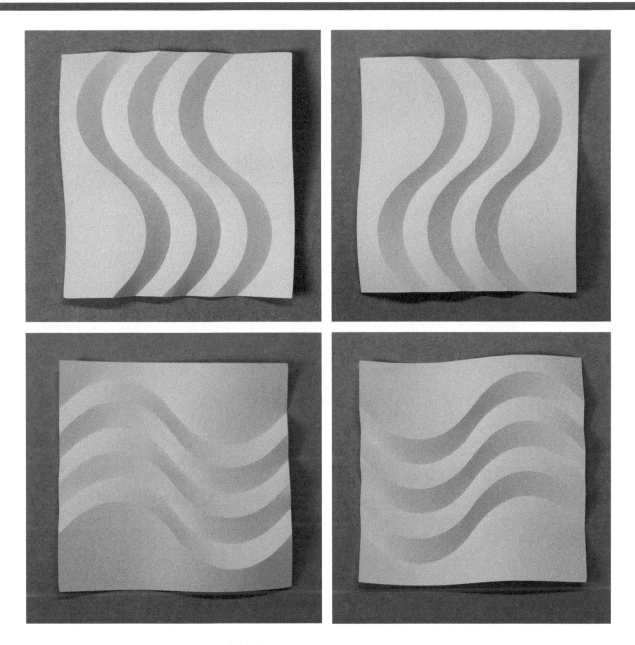

Figure 2.14   Figure 2.13 crease pattern folded.

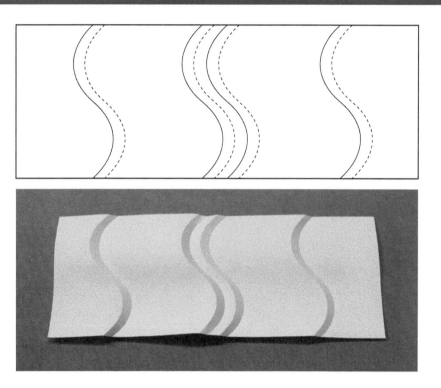

Figure 2.15   Winding curves aligned in an unequally spaced manner.

2.3 Aligning Curve and Reversed Curve

The previous examples were the same-shaped curve translated and aligned. Now, let's align a curve and its mirrored image side by side. First, align each of the curved pleat folds and its mirror image, as seen in Figure 2.16. Keep the mountain and valley as is before and after mirroring the reverse. The result is a shape with the mountain folds facing each other across

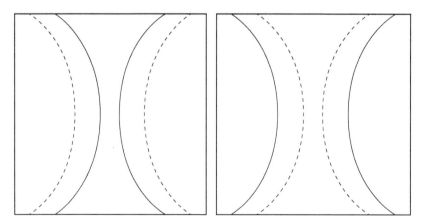

Figure 2.16   Crease pattern where multiple simple curves are aligned.

the center (left in Figure 2.17) and a shape with the valley folds facing each other across the center (right in Figure 2.17). These are the same shape viewed from front and back.

This example can also form a cylinder by folding at a large angle and rolling the whole shape. The crease pattern in Figure 2.18 is similar to the left one in Figure 2.16, except with wider spacing to prevent partial

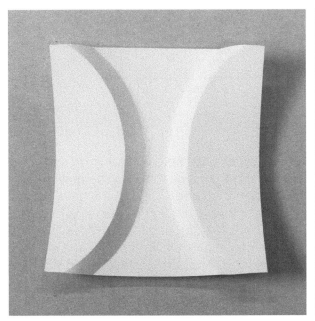

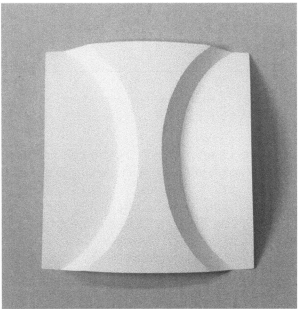

Figure 2.17　Figure 2.16 crease pattern folded.

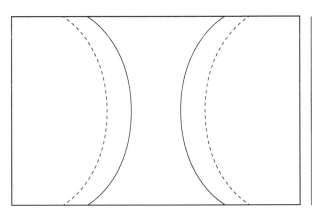

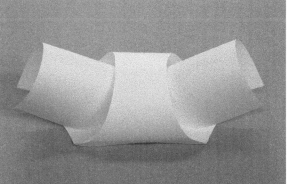

Figure 2.18　Crease pattern with widely spaced pairs of fold lines and folded shape with cylinders on both ends.

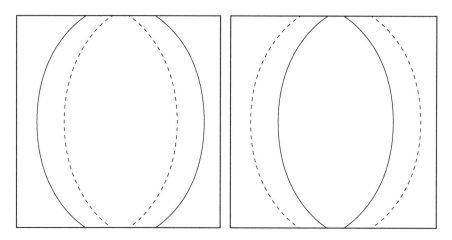

Figure 2.19   Crease patterns where concaves are facing.

interference after folding. Folding and rolling this makes an interesting shape with a cylinder projecting on both sides.

Try with the curves facing each other on the concave side. Figure 2.19 shows the crease patterns and Figure 2.20 shows their folded state. This way of shading creates a different feel to the appearance.

The crease pattern in Figure 2.21 is similar to the left one in Figure 2.19, except for the wider center part. Folding at a large angle and rolling the piece makes a shape with a cylinder in the center.

Let's try some winding curves. Figure 2.22 is an example of mountain wavy lines laid out facing each other. Figure 2.23 shows the pleat-folded shapes from the aligned wavy lines of Figure 2.22. Just translating or reversing a curve will give you beautiful shading patterns.

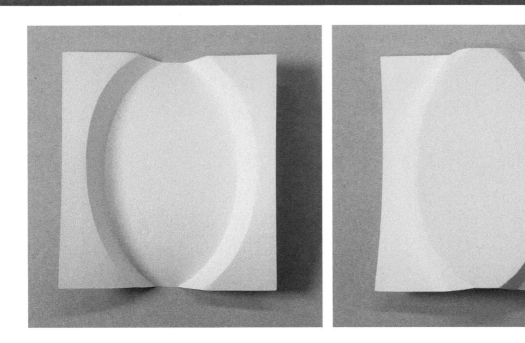

Figure 2.20   Figure 2.19 crease patterns folded.

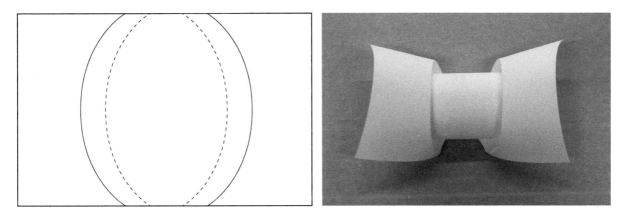

Figure 2.21   Crease pattern with widely spaced pairs of fold lines and folded shape with cylinders on both ends.

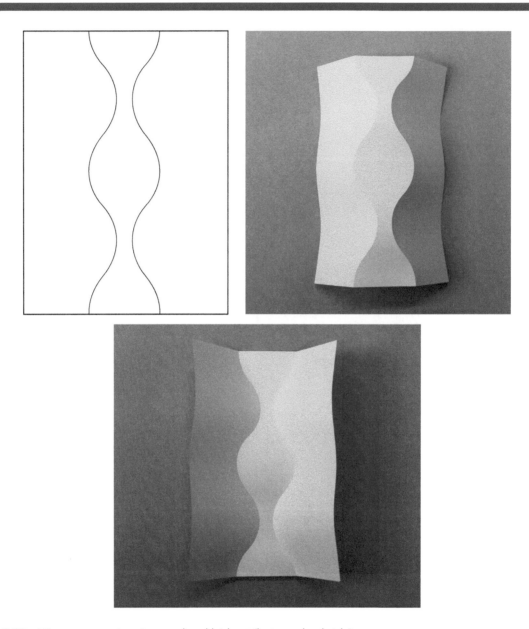

Figure 2.22   Wavy curves mirror-inverted and laid out (bottom: backside).

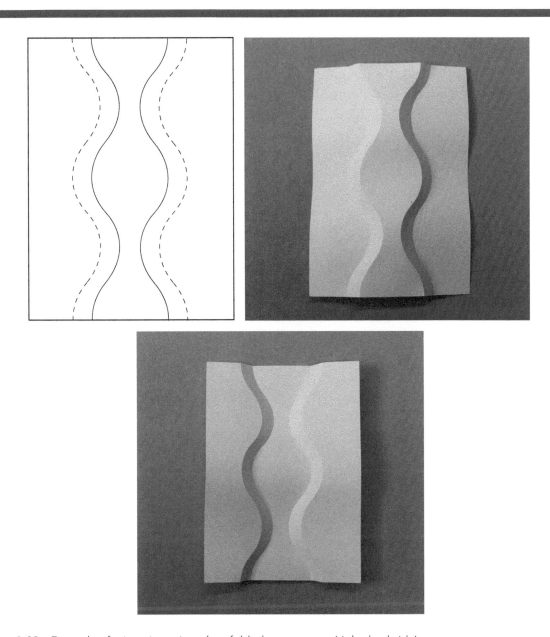

Figure 2.23   Example of mirror-inverting pleat-folded wavy curves (right: backside).

Artwork 5: Ring

This work is terraced by aligning simple curves as-is or reversed. It uses a paper strip with both ends glued into a ring (the gray portion in the crease pattern is glued). Notice that the curves alternate between mountains and valleys, except when the curve and its reverse are facing each other where both are mountain folds.

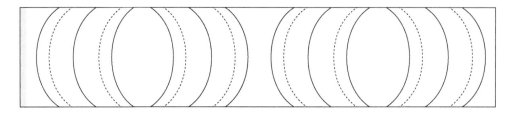

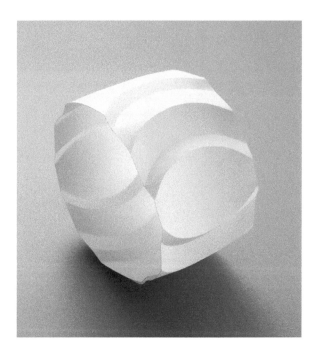

Artwork 6: Terraced Mound

A randomly made wavy curve is aligned repeatedly. Six of them are placed at certain intervals on the left, with alternating mountain and valley folds. Then, the entire left portion is reversed to make the right half. Even a simple approach like this can produce a wonderful solid with a beautiful shading.

Artwork 7: Shading of Riffles

This textured pattern starts with the pleat-folded winding squiggles, which are then reversed and repeatedly aligned. Instead of a plane parallel alignment, the squiggles are diagonally translated to add a bit of spice. The shallow recesses and projections bring elegant lines of shading to the surface.

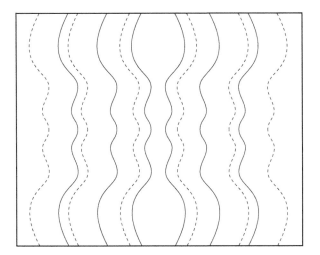

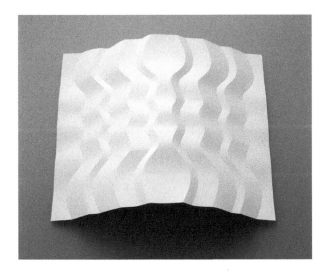

For Those Who Want More Theory–2

RELATION BETWEEN MOUNTAIN/VALLEY FOLD LINES AND PROJECTION/RECESS OF PAPER

The paper folded at a single curve will have two curved surfaces across the fold. In Figure 2.24, (same as Figure P.8), a convex curved surface emerges to the left of the fold and a concave curved surface to the right.

The state of a curved surface like this is predictable from the gradual shading with illumination, but not exactly predictable depending on the shape or the photographed state. If we can see on the crease pattern whether the curved surfaces will be convex or concave after folding, it greatly helps us design and fold shapes.

So, let's mark the crease pattern and the folded shape with a "+" sign on the projection after folding and a "−" sign on the recess, as in Figure 2.25.

The left side of the mountain fold won't be concave and will always be convex. So, the symbols shown in Figure 2.25 are not affected by the fold angle. When the mountain and valley folds are inverted (equivalent to flipping the paper), the crease pattern and the folded shape are as in Figure 2.26.

Note that the + and − symbols indicate the state near folds, not the concavo-convex state of a wider range. Figure 2.27 indicates the projection and recess for the paper pleat-folded at the winding curves. The projection and recess are seamlessly connected, with no strict border between them. The vicinities across a straight fold are planar.

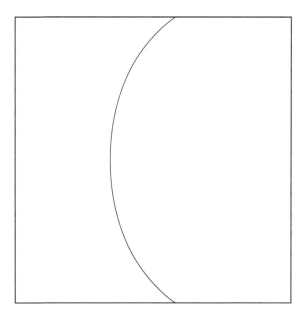 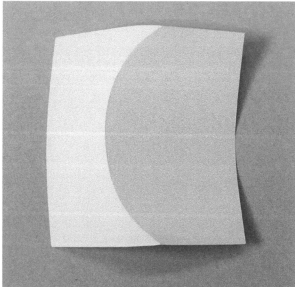

Figure 2.24   Shape from folding one simple curve (same as Figure P.8).

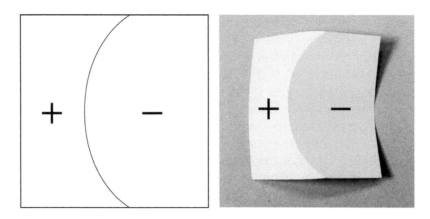

Figure 2.25   Curved surface marked with plus and minus signs on projection and recess, respectively.

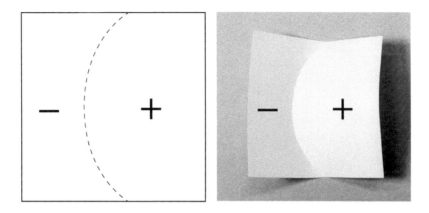

Figure 2.26   Inverted mountain and valley folds of crease pattern Figure 2.25.

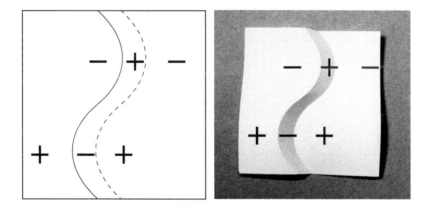

Figure 2.27   Projection and recess with symbols on wavy curved surface.

Next, several new symbols are introduced, as seen in Figure 2.28, to indicate whether the fold line is a mountain or valley and the curved surface on each side is convex or concave. The central curve symbol represents the shape of the fold line (C-shape). The mountain and valley fold lines are marked with "+" and "−" in the circle, respectively. The square contains the sign indicating whether the curved surface across the fold line is convex or concave. The sign "+" means the curved surface is convex. The sign "−" means the curved surface is concave. There are only two possible combinations of signs, as seen in Figure 2.29, like those we saw in Figures 2.25 and 2.26. No other combination exists.

Take a close look at these symbols. You'll find that the sign is the same in the circle and in the square connected to the circle. Another square has a different sign. On closer look, if one sign is determined, the remaining two are automatically decided.

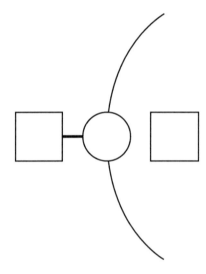

Figure 2.28   Newly introduced symbols.

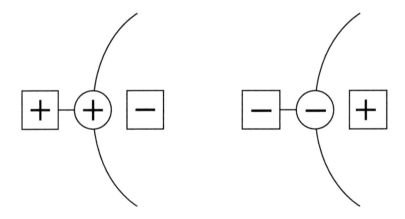

Figure 2.29   Sign entry example.

Now, let's do an easy exercise.

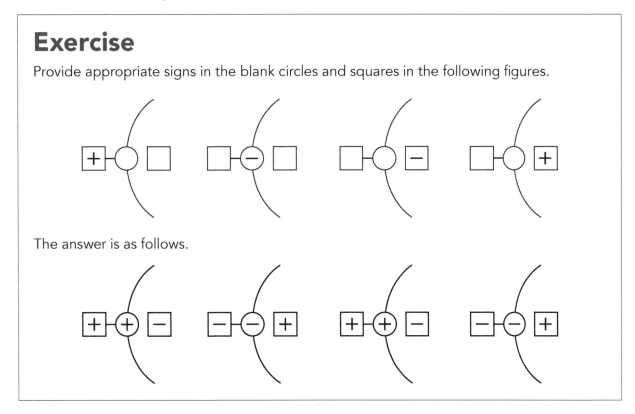
This easy rule helps you predict the projection and recess of curved surfaces after folding, based on the shape and mountain/valley assignment of the fold lines in the crease pattern. This knowledge is also effective when assigning mountains and valleys after aligning the fold lines.

Figure 2.30 is an example of where the same-shaped curves a and b are aligned in parallel. Assume that you assign the curve a to a mountain-fold (Step 1). This automatically decides whether the curved surface on each side of it is convex or concave (Step 2). The adjacent curve b shares the portion between the fold lines a and b. So, the square on the left of b has a "−" sign (Step 3). This decides the remaining signs for the curve b, making the curve b to be valley folded and the right end convex (Step 4). In summary, the same-shaped curves aligned side by side alternate between mountains and valleys. If the left one is mountain-folded, the both ends are projected and the center is recessed.

Let's take a look at another example. Figure 2.31 is an example of where the curve a and its reverse b are aligned side by side. The curve a is mountain folded (Step 1). This automatically decides whether the curved surface on each side of it is convex or concave (Step 2). The adjacent curve b shares the portion between the fold lines a and b. So, the square on the left of b has a "−" sign (Step 3). Apply here the reversed symbol set of the left one in Figure 2.29. You'll find that the curve b is a mountain fold and the right end becomes convex (Step 4). In summary, a curve and its adjacent reverse are both mountain and valley folds. The curved surface after folding is also predictable from the sign in the square.

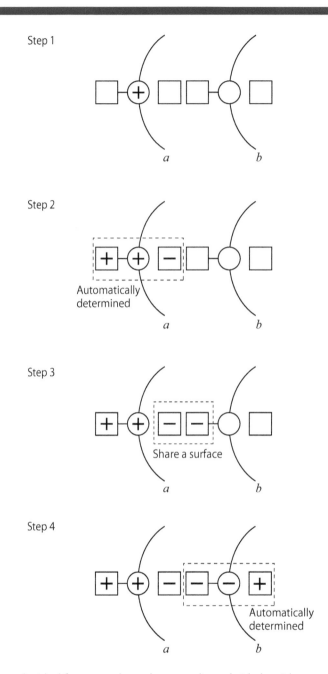

Figure 2.30   How signs are decided for same-shaped curves aligned side by side.

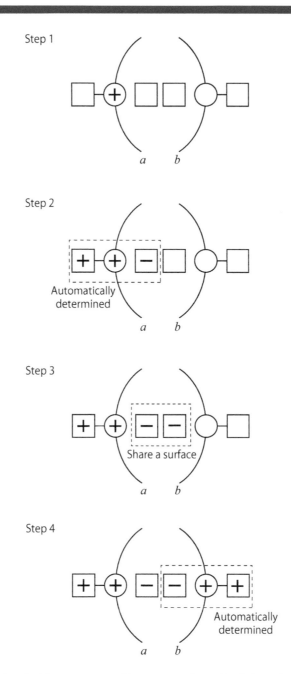

Figure 2.31   How signs are decided for a curve and its reverse aligned side by side.

Chapter 3

Aligning Rotated Curves

We have translated and aligned the same-shaped curves in parallel. In this chapter, we are going to make rotationally symmetric structures by aligning the same-shaped curves in a circular manner.

3.1 Making Rotationally Symmetric Structures

Rotate a curve at the angle in increments of given degrees and align them at their end points. In Figure 3.1, six curves are aligned at 60-degree intervals. The curves alternate between mountain and valley folds, as for the same-shaped curves aligned in parallel. When you rotate this crease pattern 120 degrees, the rotated pattern is exactly the same as the one before rotating. Shapes like this are called rotationally symmetric shapes. This crease pattern folds to a shape, as seen in Figure 3.2.

Try with a different number of curves. This number must be even, as mountains and valleys need to alternate. To lay out an even number of curves at intervals of a given degree, set the rotation angle to a certain value (like 15, 30, and 45 degrees) obtainable by dividing

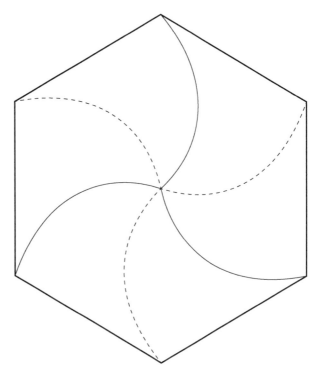

Figure 3.1 Crease pattern where rotated curves are aligned.

180 degrees by an integer. In Figure 3.3, a curve is rotated by increments of 45, 30, and 22.5 degrees to lay out a total of 8, 12, and 16

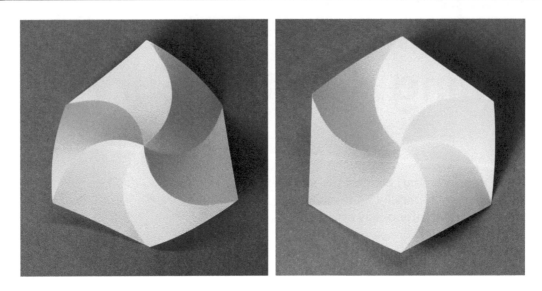

Figure 3.2   Figure 3.1 crease pattern folded. Right photo: backside.

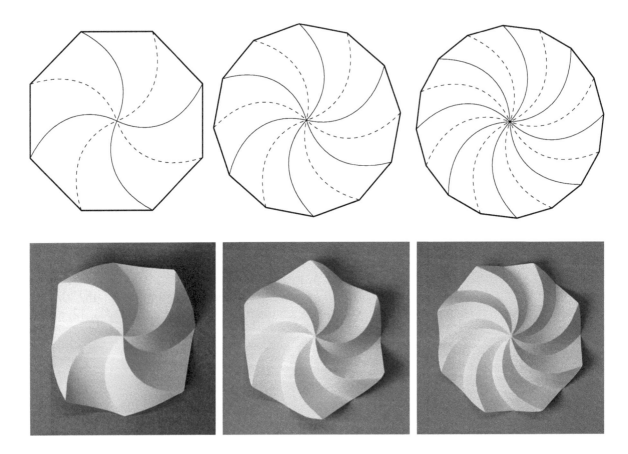

Figure 3.3   Different number of curves. 8, 12, 16 curves.

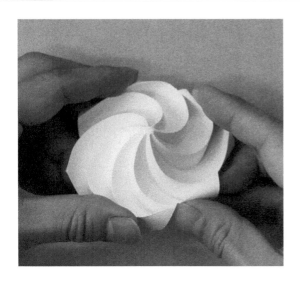

Figure 3.4 16-curve shape (Figure 3.3) tightened up.

curves rotationally symmetrical, respectively. The shape of this curve is decided intuitively, not theoretically, but still is neatly foldable. The number of fold lines affects the density of shading, creating a different feel to the appearance. Crease and release the workpiece to settle itself down. It looks like a cupcake when wadded, as shown in Figure 3.4. You may want to keep it in this state by framing or gluing.

Figure 3.5 is an example of where 12 winding curves and 16 pointed curves are aligned rotationally symmetrically. The angular curve is added with a new fold line at the angle portion, as in the example on page 8. Non-simple curves also create beautiful shapes. Try with many more curves.

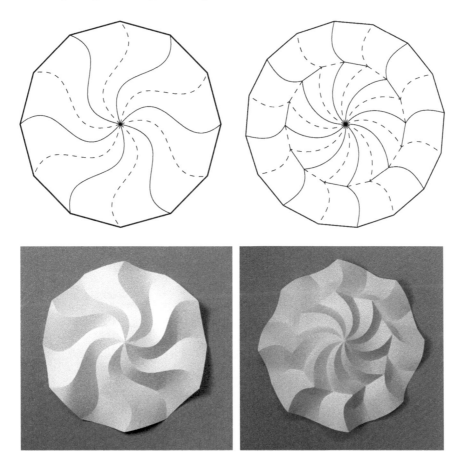

Figure 3.5 Winding and angular curves formed into rotationally symmetrical structures.

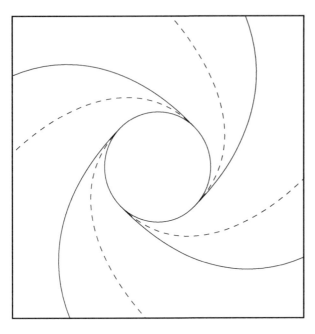

Figure 3.6   Crease pattern with a circle in the center.

3.2 Decentering

Let's move the fold lines away from the center point. As in Figure 3.6, lay out the curves so that they touch a circle placed in the center. Each pair of mountain and valley folds should touch the circle at the same position. The folded shape has a lineless blank portion in the center, as seen in Figure 3.7, creating a different feel to the appearance. When folded more tightly, the result looks like a disk reeling the paper.

3.3 Aligning Whirls

With some elaborations, crease patterns with spiral fold lines can be aligned and connected. In Figure 3.8 crease pattern, the curve is reshaped so that the fold line is perpendicular to the contour of the paper. When the entire

Figure 3.7   Curves folded after moving from the center: Right: backside.

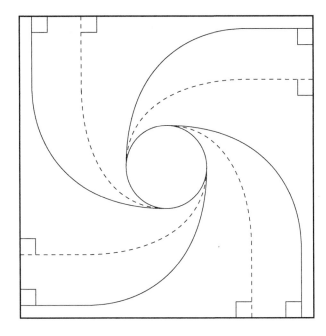

Figure 3.8   Crease pattern where fold lines are perpendicular to contours of the paper.

pattern is reversed and aligned side by side with the original, the fold lines are connected seamlessly to make the crease pattern, as seen in Figure 3.9. Figure 3.10 is obtained by reversing Figure 3.9 crease pattern and

connecting it vertically to the original. In this way, a four-whirl pattern can be folded out of a sheet of paper.

Align Figure 3.10 crease patterns with as many as you like to extend. An origami work tiled with the same shapes (or units) is called "tessellation". A variety of tessellation works have been created by many origami artists. Figure 3.11 shows two types of diagrams describing how whorl units should be laid out to connect to each other. The unit's center comes at the grid's intersection. At this point, the unit's rotation direction should be opposite to the direction of the destination unit (the arrow in the diagram indicates the rotation direction; the grid point is color-coded to show the direction). In the left diagram, where units are laid out in square grids, four pairs of mountain and valley folds are made to connect to four surrounding units. In the right diagram, where units are laid out in regular hexagonal grids, three pairs of mountain and valley folds are made to connect to three surrounding units.

The sample works on pages 41 and 42 use the units with four and three mountain-valley fold pairs aligned in square and regular hexagonal grids, respectively.

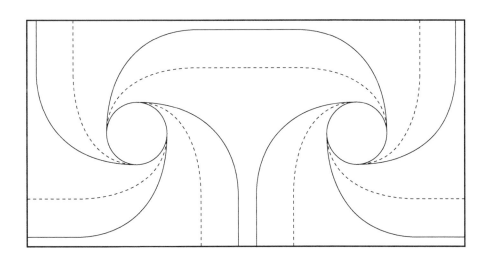

Figure 3.9   Whirl structure connected to its reverse.

Figure 3.10   Four whirl structures connected.

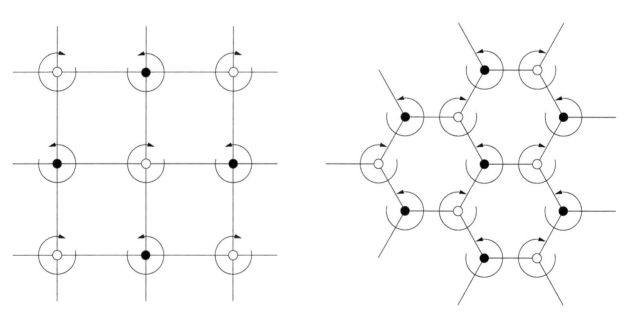

Figure 3.11   Whorl unit connection patterns.

Artwork 8: Screw

Five mountain-and-valley fold pairs are aligned rotationally symmetrically. The angles made from pairs of 10 fold lines in total are not equally spaced, but alternate at large and small degrees. The deeply pushed center part makes the shape three dimensional. The shape is fixed by folding corners back partially.

Artwork 9: Whirl Tiling

The 8-fold-line pattern (four mountain-valley pairs) in Figure 3.3 is laid out in the regular square grids seen in the left diagram of Figure 3.11. This work has a total of 16 whirls. Mountain fold lines smoothly connect to the adjacent whirls but valley lines collide halfway. The straight lines inserted in between them deliver a beautifully indented pattern.

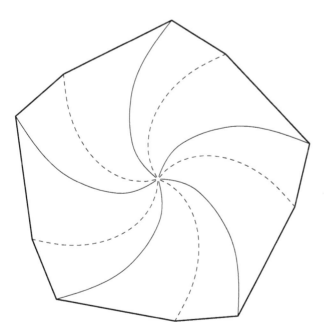

Artwork 10: Tri-Wing Boomerang Tiling

A pattern of neatly aligned tri-wing boomerangs emerges from pairs of rotationally symmetrical fold lines aligned in grid. Notice the points at which the fold line ends gather on the crease pattern. Each of these points is the center for three steps created from pairs of mountain and valley folds laid out at 120-degree intervals. These rotationally symmetrical fold lines are laid out in the regular hexagon grids, as seen in the right diagram of Figure 3.11. The result is an alignment of popped up boomerangs.

Appendix: Is Paper Really Inelastic?

Paper is said to be an inelastic material. So, a shape from paper can be mathematically expressed as a set of developable surfaces. However, is it true that paper does not expand or contract? Usually, paper thickness is ignored in mathematical expressions. However, real paper has thickness. Think about the cross section of a sheet of paper folded twice, as shown in Figure 3.12. Paper cannot be folded into this state if it is absolutely not stretchable. Ideally, the fold width should be zero. Actually, paper expands and contracts near the fold little for its thickness.

Paper is made from fibers, which get broken or come loose at the folded portion to cause a minute expansion and contraction. If you press some soft-ish paper to a curved surface and rub the paper, the rubbed portion pops up. This is used in embossing. At the point that is concentrated with folds (Figure 3.13), the real thing after folding greatly differs from the theoretical shape. Real shapes are not describable by simple equations. This is a challenging problem when designing but is an expressive power on shapes made from folding paper.

While folding paper, I often hit on unexpected shapes that are computationally impossible. The material property of paper greatly widens the possibility of origami.

Figure 3.12   Cross section of folded paper.

Figure 3.13   Point concentrated with fold lines.

Chapter 4

Tucking

The previous examples contain no crossing or branching curves. In this chapter, understanding the tucking mechanism allows you to make shapes using layouts of branching fold lines.

4.1 Tucking One Line

Inside-reverse-fold, a flat-folding origami technique, is to push down a part of the folded fold line inside, as shown at the top of Figure 4.1. This is often used to make a beak of an origami bird. The origami crane also uses this technique at the last step.

Let's see how the crease pattern changes before and after the inside-reverse-fold. The single mountain fold line branches into three at the point where the inside-reverse-fold starts. The newly branched two outer lines are mountain folds. The original fold line changes from mountain to valley fold after the branch point. Figure 4.2 shows how the real paper is inside-reverse-folded. Viewing from the top, one mountain fold branches into three lines of mountain, valley, and mountain. From the backside, one valley fold branches into three fold lines of valley, mountain, and valley. If the original fold line is a mountain, the Y-shape consists

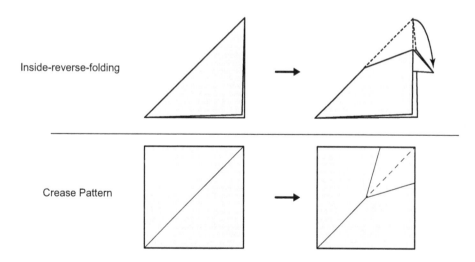

Figure 4.1   Inside-reverse fold.

Figure 4.2   Inside-reverse-folded (right: backside).

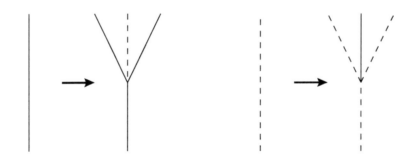

Figure 4.3   Change in crease pattern after inside-reverse fold.

of mountains emerging after the inside-reverse-fold. If it is a valley, the emerging Y-shape consists of valleys, as illustrated in Figure 4.3.

This manipulation can also be performed for curved-fold lines. Figure 4.4 is a crease pattern newly designed by applying the inside-reverse-fold technique to a simple curved-folding. Folding this crease pattern produces the left shape in Figure 4.5. The mountain folded curve changes to valley half-way, producing a recess above the valley line. It adds beautiful shading without wrinkling the paper. The right is the backside of the folded paper.

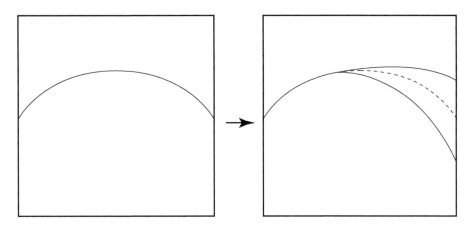

Figure 4.4 Example of applying inside-reverse fold to a simple curve.

Figure 4.5   Simple curve tucked (right: backside).

4.2 Tucking Multiple Curves

Tucking of a part of the paper, like the inside-reverse-fold, is applicable also to multiple curves on a crease pattern. Figure 4.6 is an example of tucking three aligned pieces of a curve. Add a branching fold line without changing the shape of the original curve. Reverse the mountain and valley after the branch point. Now, new folds are added to the curved-pleat-folds (Figure 4.7).

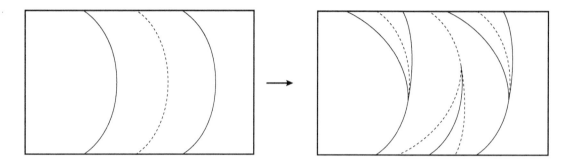

Figure 4.6 Change to crease pattern where multiple curves are tucked.

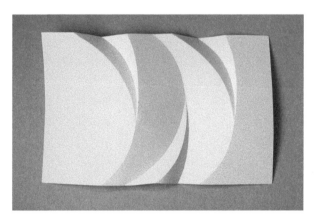

Figure 4.7 Multiple curves tucked.

4.3 Repeating Tucking

Tucking needs two additional fold lines for each fold line. The additional fold lines can further be tucked. By repeating tucking, a fold line branches into lines, which further branch into more lines, as shown in Figure 4.8 tree diagram.

The tips after branching have alternate mountains and valleys from the outermost line. Even for an uneven branching, as in Figure 4.9, the mountain and valley folds appear alternately from the outermost line. The knowledge of this rule is useful for first deciding the fold lines and then assigning mountains and valleys. Figure 4.10 is an example of branching from a curved-fold line.

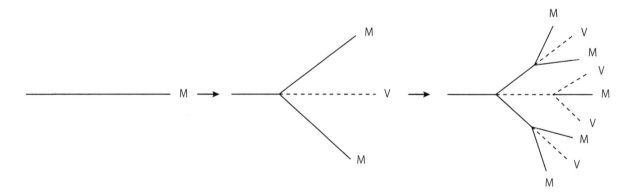

Figure 4.8 How a fold line changes with additional tucks.

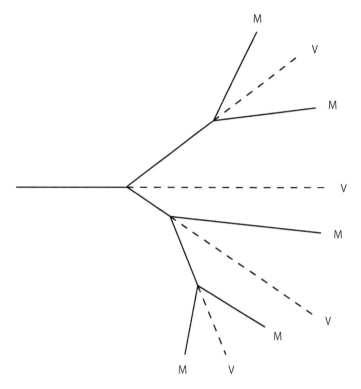

Figure 4.9 Mountain-valley assignment for unevenly branching tucks.

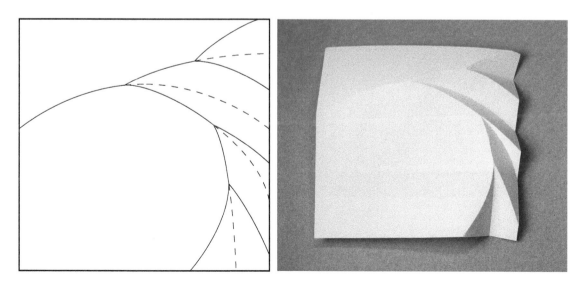

Figure 4.10 Repeated tucks on a curve.

Artwork 11: Mountain Range

A tuck-based branch is repeated as many times as necessary. The result is this mountain range-like shape. The fold lines alternate between mountains and valleys at the edge of the paper.

written freehand with a ballpoint pen. The jagged line produced a rugged feel to the shading. To finish it right, remember that mountain and valley folds should alternate at the edge of the paper.

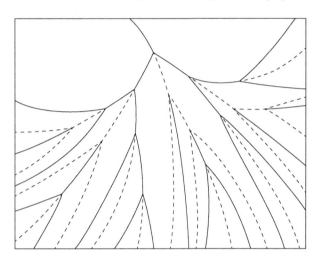

Artwork 12: Freehand-Based Mountain Range

Similar to the previous work, this one has many tuck-based branches. The base line is

Artwork 13: Rotating Star

The fold line branch technique is applied to the rotationally symmetrical fold line pattern presented in Chapter 3. The result is this popping star consisting of curves.

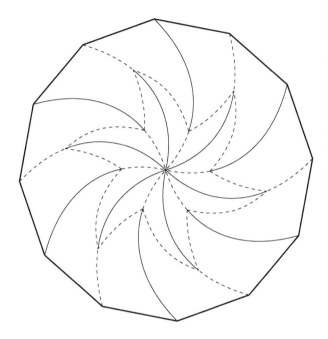

For Those Who Want More Theory–3

Can Fold Lines Intersect Freely?

In this chapter, we have looked into how a fold line branches into other lines by "tucking". From a different point of view, the fold lines intersect at one point. Can fold lines intersect freely? Unfortunately, they cannot. That's one of the difficulties in designing curved-fold origami.

Now, let's think of a situation where two fold lines intersect, as seen in Figure 4.11. In the example (a) two mountain-fold curves are intersecting. In Figure 4.11(b) one mountain-fold curve is intersecting with one valley-fold curve. Are these examples foldable with no trouble? You will try and find it impossible. Why is it?

Let's check the reason for not folding well. First, assume that fourfold lines are extending from the intersection. Provide each line with the symbols in "For those who want more theory–2" on page 28.

The mountain-fold line should be marked with

and the valley fold line with

The results are shown in Figure 4.12. The lines divide the paper into four areas. In the area marked with "?", a projection and a recess are mixed. This area does not have a defined curved surface, leaving it to not folding well.

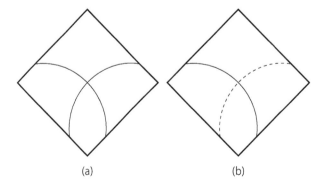

Figure 4.11   Example of two curves intersecting.

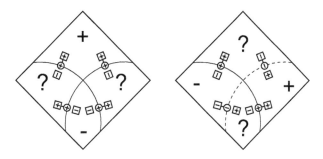

Figure 4.12   Four curves marked with mountain-valley and convex-concave symbols.

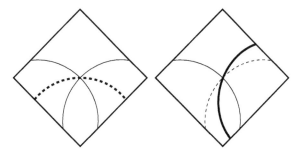

Figure 4.13   New curve (bold line) added.

So what should we do? One solution is to add a new fold line, as shown in Figure 4.13. Then, place the symbols indicating the fold line and curved surface's projection and recess.

The result is seen in Figure 4.14, where the projection and recess are defined in every area surrounded by the fold lines. For visibility, the figure color codes the projected area ("+" sign) in gray and the recessed area ("−" sign) in white. These crease patterns are neatly folded as photographed. Again, adding a new fold line is a solution for intersecting curves that are not folded well.

By the way, I explained on page 8 that it is a good idea to add a straight fold at a place bending at an acute angle. For Figure 4.15, the crease pattern on the left top of Figure 4.14, the bold mountain fold line is regarded as having an acute bend. Considering this, adding a straight-valley-fold here will create a sharper shape. Figure 4.16 is an example of adding a valley-fold line.

Figure 4.17 sample work is produced from repeating this crease pattern four times and aligning them. With some fold lines added outside, the whole piece is finished beautifully. You will create new works by observing the layout of mountain and valley fold lines on the crease pattern and combining them well.

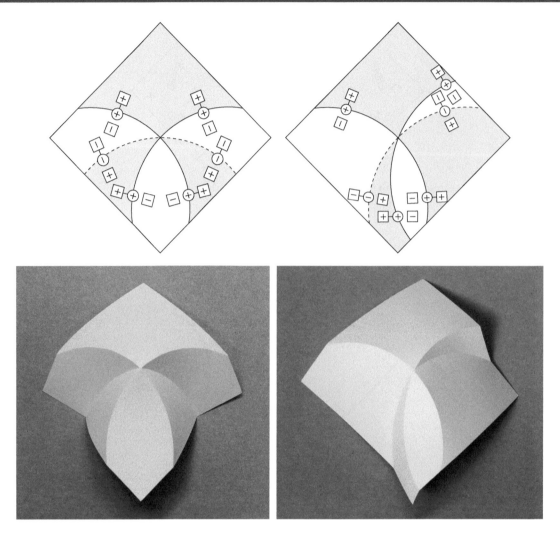

Figure 4.14   Symbol-based projection/recess judgment and folded shapes.

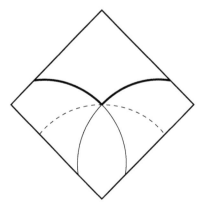

Figure 4.15   Bold line regarded as a mountain-fold line with an acute angle.

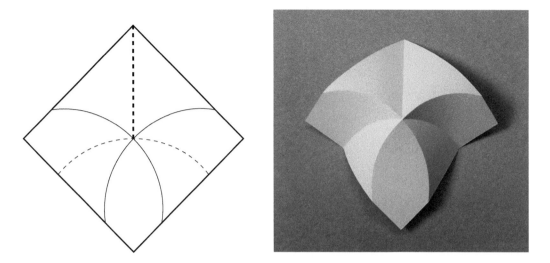

Figure 4.16   Crease pattern with an additional new straight-valley fold (bold) and folded shape.

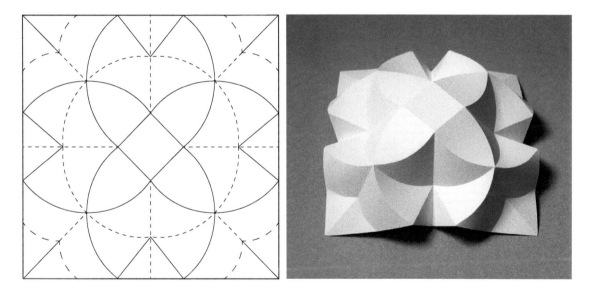

Figure 4.17   Work from combining four crease patterns of Figure 4.16.

Chapter 5

Folding Cones

Cones are one of the curved surfaces creatable with paper. A cone has its line elements concentrated to one point. This chapter presents how to fold and utilize shapes that have a cone.

5.1 Making a Cone without Cut

First, take a look at the development of a general cone. A cone is created from a sector of a circle, as shown in Figure 5.1. Cut out a sector and roll it up so that the angle becomes the tip of the cone. Paste two straight sides to form a cone as in the photo. The central angle of the sector determines the cone's vertex angle (at the tip). A sector-shaped cutout quickly makes a cone.

Alternatively, lay out one each of the mountain- and valley-fold lines on the square paper to make a conical shape, as shown in Figure 5.2 crease pattern. Fold the mountain- and valley-fold lines at 180 degrees. Overlap the entire gray areas with each other to make the center of the paper (end of the fold lines)

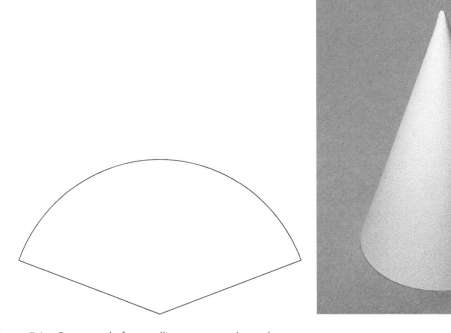

Figure 5.1   Cone made from rolling a sector-shaped cutout.

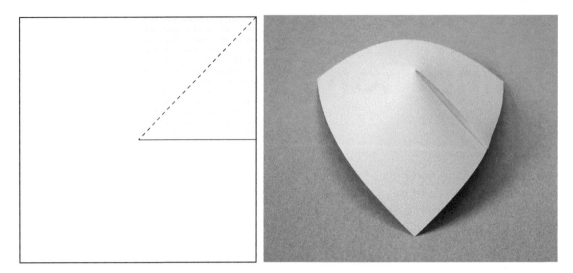

Figure 5.2 Making a cone without cutting out a sector shape.

the vertex of the cone. The cone's vertex is adjusted by changing the angle made by the mountain- and valley-fold lines on the crease pattern.

Next, add a new fold to this cone. Now, add the crease pattern with a circle centering around the cone vertex (paper's center), as seen in Figure 5.3. This folds into a shape with the cone's tip pushed inside.

The pushed-in portion in Figure 5.3 is also conical, and thus can be folded back upward by adding a circular fold line.

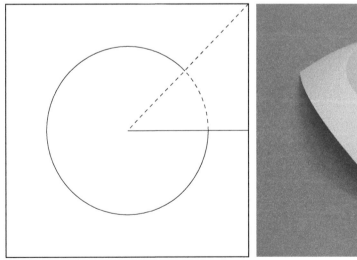
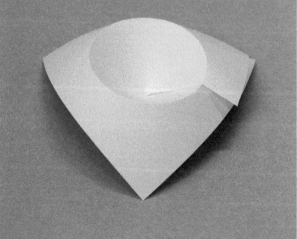

Figure 5.3 Folding back a cone.

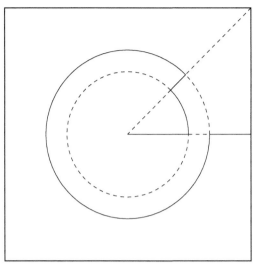
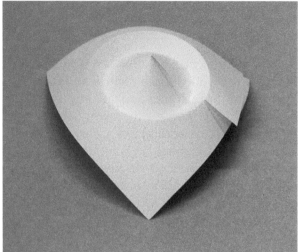

Figure 5.4   Pleat-folded cone.

Figure 5.4 crease pattern with an additional circle folds into a pleat-folded cone. Once you understand the principle, you may add more circles to your crease pattern to pleat-fold the tip many more times.

The crease pattern, consisting of four rotationally-aligned pieces of Figure 5.4 crease pattern, will fold into an interesting shape, as shown in Figure 5.5.

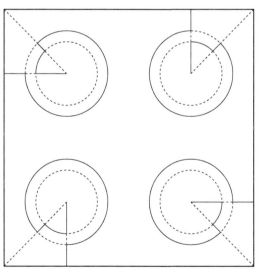
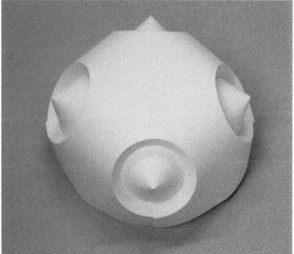

Figure 5.5   Four Figure 5.4 crease patterns connected.

5.2 Making Conjoined Cones

A 60-degree vertex angled cone is made from the sector with a 180-degree central angle (i.e., semicircle), as in Figure 5.6. Joining two copies of this pattern makes the central angle 360 degrees. So, just adding one valley-fold line on a square sheet, as in Figure 5.7, produces two conjoined 60-degree vertex angled cones. The square's center becomes the cone's vertex, making a cone on each side across the valley-fold line.

Using a similar idea, two orthogonal valley-fold lines on a square sheet in Figure 5.8 produce four conjoined cones. Each of the four areas separated by valley-fold lines becomes one cone.

You do not have to make beautiful cones. Try with variations of flabby surfaces with different positions or numbers of valley-fold lines. You'll find unexpected shapes. The resulting shape is an assembly of cones, which can be pushed in or pleat-folded by adding fold lines as presented in the previous section.

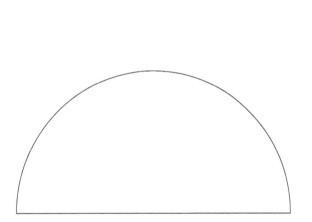

Figure 5.6   A 180-degree central angled sector makes a 60-degree vertex angled cone.

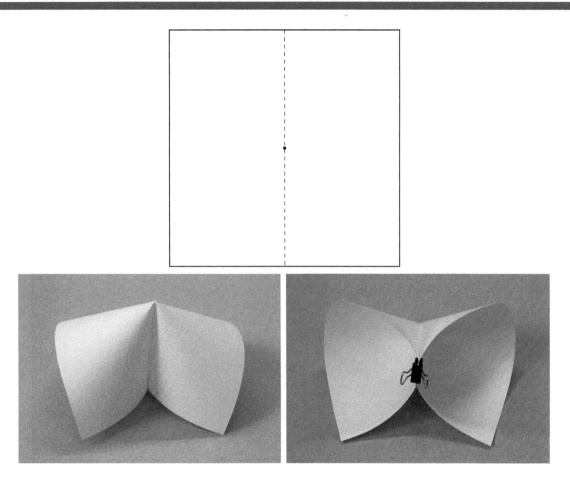

Figure 5.7   Two conjoined cones made without cutting the paper.

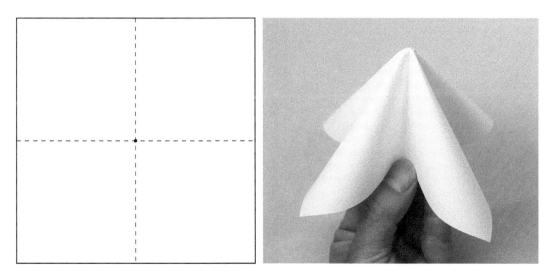

Figure 5.8   Four conjoined cones made without cutting the paper.

5.3 Making a Partial Cone

We have made various cone-based shapes until now. However, it is difficult to combine a cone with other fold lines. So, let's try this approach: fold a cone partially and connect it to a different fold line.

Figure 5.9 crease pattern, using a combination of an arc and lines, will produce a shape containing a part of the cone. After folding, the inside of the arc becomes conic. Let's combine these to produce new shapes.

For instance, Figure 5.9 crease pattern makes a conical hollow. Laying out this crease pattern at each side of a square sheet produces Figure 5.10 shape.

Connecting the arc-fold lines makes a wave-like fold line, as shown in Figure 5.11. This looks like Figure 1.13 example (page 6), but has a clearer indented surface due to the straight-valley-fold line passing through the center of the arc. In Figure 5.12, each partial cone is further separated by

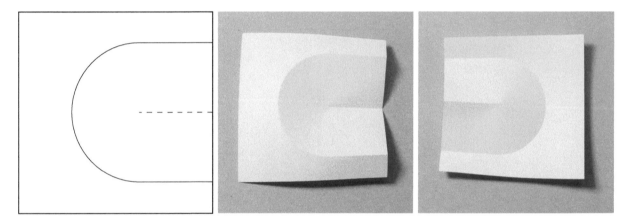

Figure 5.9   Crease pattern to fold a part of the cone and the resulting shape (right: backside).

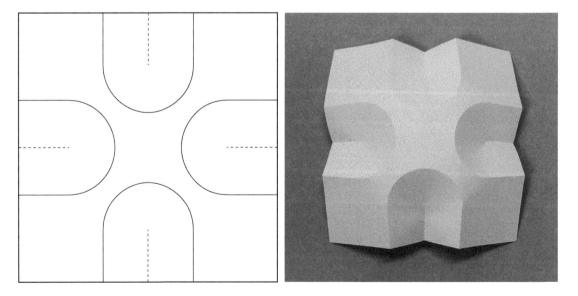

Figure 5.10   Partial cone laid out at four surrounding sides.

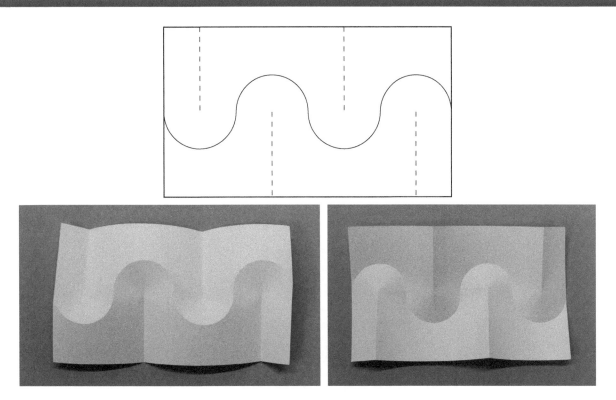

Figure 5.11   Identical partial cones connected into a wave-like fold line (right: backside).

Figure 5.12   Straight line inserted in Figure 5.11 example.

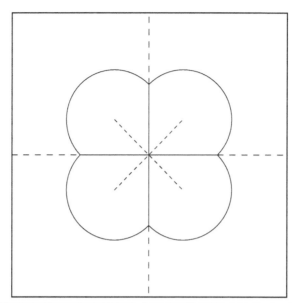

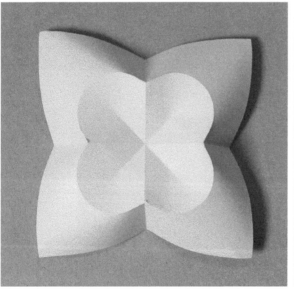

Figure 5.13   Four identical partial cones laid out facing the center.

a straight line inserted when connecting the part of the cone.

In Figure 5.13, four tilted identical partial cones are aligned facing the center. This forms a flower-like pattern with a hollow center. Folding at a larger degree makes the shape more three-dimensional. This may be equivalent to the four conjoined cones in Figure 5.8, whose tips are pushed in.

Artwork 14: Flower

Four identical partial cones are laid out toward the center, and their curves are connected into a whirl. Folding at a large angle formed this flower-like shape. The shape is stabilized by fixing at the portion where the paper surfaces meet.

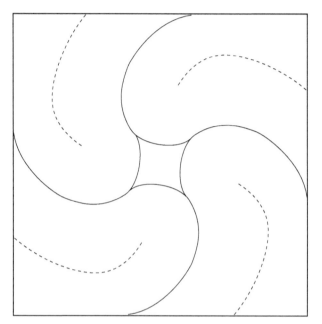

Artwork 15: Japanese Rock Garden 1

This work is made by combining the techniques, such as partial cones, branching fold lines, and curved-pleat folds. The shading looks like ripple marks in a Zen garden.

Appendix: Folding Back a Cone

As seen in this chapter, a cone tip can be pushed in then partially pushed up to pleat-fold, just by adding fold lines, without cutting the cone. The cone tip is not necessarily pushed in straight downward. No problem if it is tilted. Obliquity creates interesting shapes. Figure 5.14 shape is made from a cone tip repeatedly mirror-inverted around an oblique plane. It is difficult to manually determine the exact fold lines. So, I used a computer to obtain the line intersection between the cone and the plane.

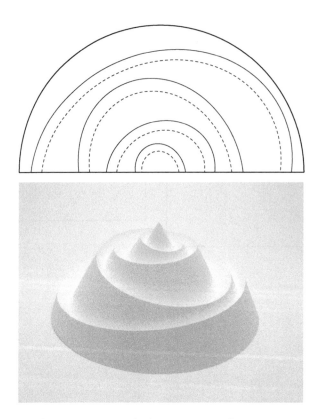

Figure 5.14   Modeling from a 60-degree vertex angled cone repeatedly mirror-inverted around the oblique plane.

For Those Who Want More Theory–4

FLAT-FOLDED ORIGAMI VERSUS 3D-FOLDED ORIGAMI

This book is intended for curved folding of paper. Regarding straight-fold origami, many studies have been done and have produced lots of knowledge. For instance, it is known that the following conditions have to be met for flat folding at the point where fold lines intersect:

- The number of mountain- and valley-fold lines differs by 2 (Maekawa's theorem)
- The alternating sum of angles made by adjacent two fold lines is 180 degrees (Kawasaki's theorem)

Figure 5.15 illustrates the above two conditions. If your crease pattern does not meet either one condition, then it is not flat foldable.

On the other hand, this rule does not apply to 3D-folded origami. For instance, Figure 5.16 crease pattern does not meet both Maekawa and Kawasaki theorems, but folds to a three-dimensional shape as in the photo, because its fold angle is not limited to 180 degrees. Figure 5.17 is a crease pattern with only valley-fold lines. This crease pattern may be folded three-dimensionally, but not without bending the paper. However, if bending is okay, a shape with four conjoined cones can be made as on page 61. Any combination of fold lines is foldable if paper bending is allowed. Still, this is only possible around the portion where fold lines intersect. Crease patterns, where multiple fold lines interrelate, are often not well foldable without wrinkling. If paper is allowed to bend, it becomes very difficult to judge whether a crease pattern is foldable (neatly without wrinkling) just by seeing it.

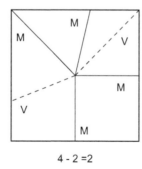

4 - 2 = 2

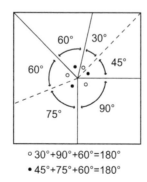

∘ 30°+90°+60°=180°
• 45°+75°+60°=180°

Figure 5.15   Explanation of Maekawa's theorem (left) and Kawasaki's theorem (right).

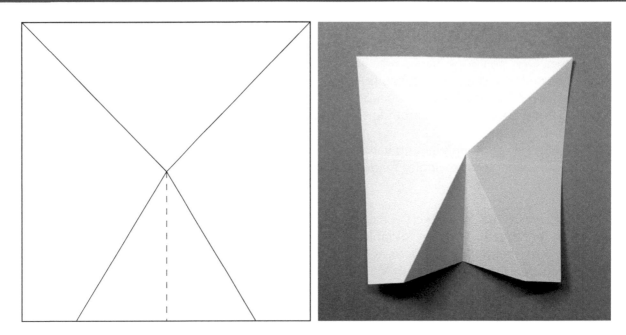

Figure 5.16 Crease pattern not meeting Maekawa and Kawasaki theorems can also be folded three-dimensionally.

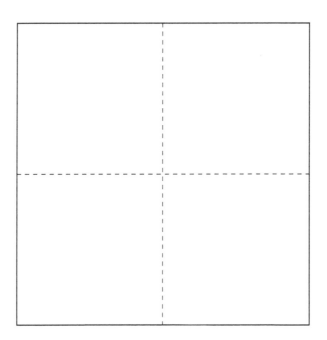

Figure 5.17 Crease pattern having only valley-fold lines.

Chapter 6

Folding Back at Straight Line

We have worked with examples using only curves. Now, adding straight-fold lines with folding back provides pleats. This creates bent pipe-like shapes unrealizable just with curves.

6.1 Combining Straight and Curved-Fold Lines

We have seen curved folds so far. Let's review straight folds now. Figure 6.1 top diagrams show valley folding at a straight line. As illustrated, the straight fold always remains linear

during folding and cannot be bent. However, after folding it back and flat at 180 degrees and bending the whole piece together, the straight fold itself is bent, as seen in the bottom center diagram. Then, adding another folding forms the shape, as seen in the bottom right diagram.

Figure 6.2 shows a layout where a straight-valley fold and a curve-mountain fold are aligned. Give it a try. You'll find that the straight-fold line has to be folded before the curved-fold line, and the straight-fold line always folds back by 180 degrees to produce a tight overlap on the paper, as explained in

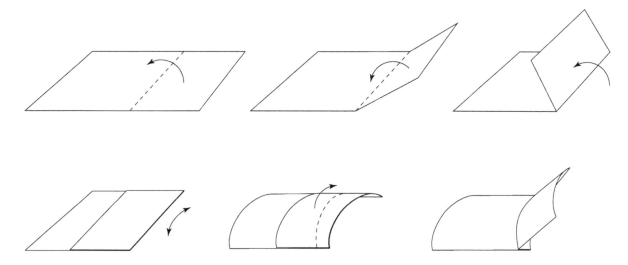

Figure 6.1   Combination of straight and curved folds.

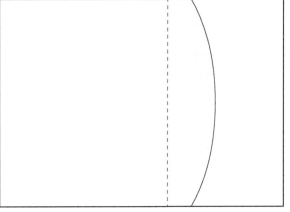
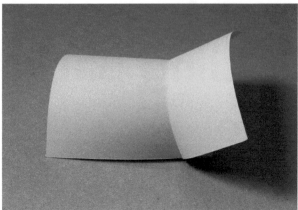

Figure 6.2   Combination of straight and curved lines (1).

Figure 6.1. The gray area in the crease pattern will be overlapped. This forms a shape with a projected surface on each side across the fold line, as in the photo. In the previous examples, if one side is convex across the curve, then another is always concave. By folding (or folding back) at a straight line, both sides across the fold are projected, forming a cylinder pipe-connected shape.

In this photo, the right side is rising at a slant. The rise angle varies depending on how much degree the curved surface is rolled. The more tightly rolled to form a cylinder, the more level the whole piece becomes. If it is less tightly rolled to be flatter, the more the right-side rise stands toward a right angle.

Figure 6.3 has a different fold line layout where a straight line is placed outside the curve (right to the curve in the figure). The overlapped portion is folded outwardly like a pleat. Similar to the previous example, the degree of rolling of the curved surface affects how the entire piece bends. The more tightly rolled to form a cylindrical shape, the less the whole piece bends and becomes more level. The less tightly rolled, the more the whole piece bends.

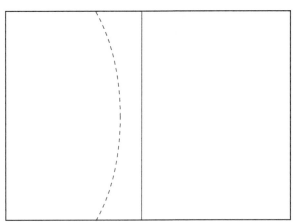
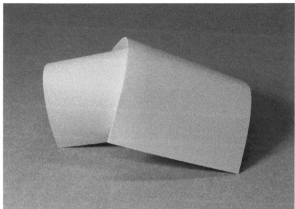

Figure 6.3   Combination of straight and curved lines (2).

70   Folding Back at Straight Line

6.2 Aligning Straight and Curved Lines Alternately

Figure 6.4 is a bent pipe-like three-dimensional shape from aligning Figure 6.1 pairs of a line and a curve. Figure 6.5 is Figure 6.2 pattern repeatedly aligned, which produces a bent cylinder shape with pleats folded outward. It looks more like, say, the back of a woodlouse. Both shapes are partially fixed to stabilize. The degree of rolling the portion that resembles a cylinder affects how the entire piece bends.

Figure 6.6 has a bit more elaborated alignment of line-and-curve pairs. In this crease pattern, all curves are identical in shape, but some of them are reversed. The gray areas will be tightly overlapped. Notice that both sides of the crease pattern are structured the same as the end of the pillowcase. Extra glue tabs are provided on the top and bottom for stabilizing the shape. This folds to a shape with a pillowcase standing upright on both sides, as in the photo. By gluing and closing the tabs, the curved surfaces intersect perpendicular to each other.

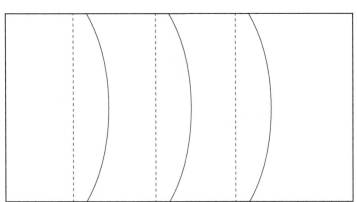

Figure 6.4   Aligning straight and curved lines alternately (1).

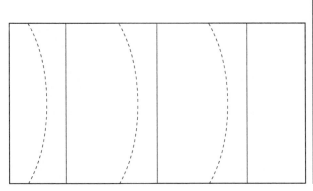

Figure 6.5   Aligning straight and curved lines alternately (2).

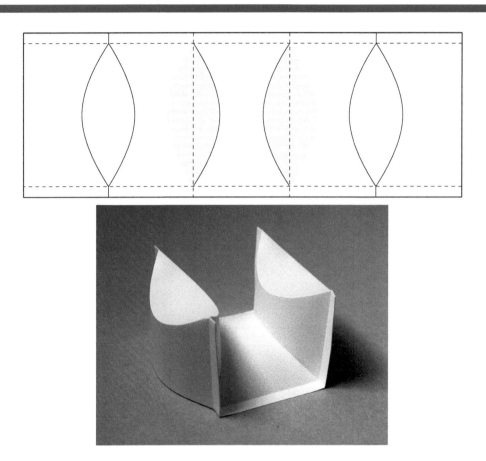

Figure 6.6   Example of upright bend.

6.3 Aligning Line-and-Curve Pairs Rotationally Symmetrically

Axisymmetric solids with external pleats are produced by aligning pairs of a line and a curve rotationally symmetrically. Fold lines do not have to meet in the center of the paper. Instead, you may place a regular polygon and extend the fold lines from it.

Figure 6.7 crease pattern has a regular decagon in the center, with a line and a curve extending from each vertex of it. This crease pattern folds to a flower shape with its folds a bit open. Using the same crease pattern, fold the straight fold lines at 180 degrees, then you'll have tightly overlapping portions. This forms a bowl, as shown in Figure 6.8. The gray portions in Figure 6.8 will be overlapped and glued to fix the shape.

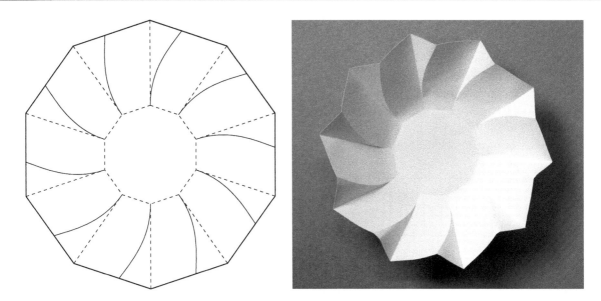

Figure 6.7   Example of line-and-curve pairs aligned rotationally symmetrically.

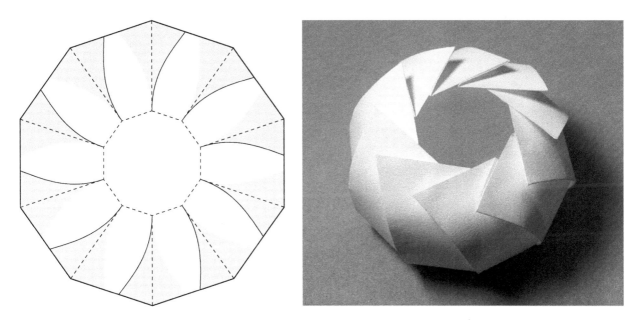

Figure 6.8   Gray areas tightly overlapped after folds.

Artwork 16: Sphere

This sphere is formed with 16 line-and-curve pairs aligned in parallel. The overlapped top and bottom portions cause both ends to tightly close after rolling. Curves drawn with hands do not work well. The crease pattern of this work was computer calculated.

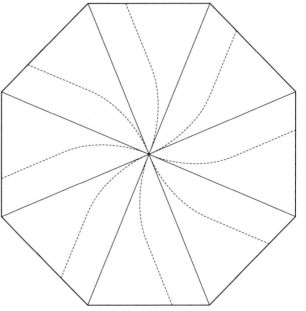

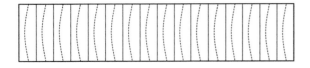

Artwork 17: Egg Wrapping

The radially placed line-and-curve pairs formed this egg wrapping shape after folds. At the base, the paper overlaps and closes tightly. This crease pattern is also computer calculated, as freehand drawing does not work out.

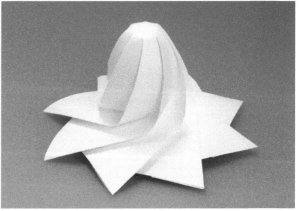

For Those Who Want More Theory–5

Folding Back a Curved Surface with Mirror Inversion

A cone was given as an example of fold back on page 58. But other shapes can also be folded back in the same way. As illustrated in Figure 6.9, (a) fold the paper into a shape, (b) put a plane crossing a part of the shape, then (c) mirror invert one portion on the plane. The shape thus obtained can be made just by folding a sheet of paper without cutting. By repeating this operation, you may create a complex shape from a simple one.

After folding back, new folds are added to the intersection of solid and plane. The folds are curves on the plane. The crease pattern is added with new fold lines, but other fold lines remain unchanged in shape. However, the mountains and valleys change for the line pushed in by mirror inversion, as in "tucking" in Chapter 4.

In principle, this is an easy operation. But you will need a computer to get accurate fold lines to prepare your crease pattern for determining the solid-and-plane intersection and mapping it on the crease pattern. Knowing this method, you may take another approach. Fold the real paper, check the shape, then trace the fold lines into the crease pattern.

Figure 6.9   Mirror inversion on crossing plane.

Chapter 7

Other Techniques

The last chapter of this book presents many other ways of laying out fold lines for various structures together with the resulting shapes. Connecting or changing these shapes helps you create a variety of shapes.

7.1 Providing Projections and Hollows

Same-shaped curves need to alternate between mountain and valley when aligned. However, a less bent, almost linear curve can break this rule. Aligning in a sequence of valley, mountain, mountain, and valley, as in Figure 7.1, makes a projection on the surface. On the backside, it looks like a groove. Figure 7.2 is the indented pattern folded from these projections and grooves aligned.

Fold lines can meet at the center of the paper, allowing them to be freely laid out without affecting each other. Grooves like this may replace the tucking explained in Chapter 4. In Figure 7.3, the fold lines are arranged so that they extend radially from the center.

7.2 Folding Cylinders

Just aligning the same-shaped curves makes a shape with a cylinder in the center, as presented in Chapter 2. Figure 7.4 shows the crease pattern and its resulting shape in Figure 2.18, rotated by 90 degrees. Horizontally expand this crease pattern and add some horizontal fold lines to make a cylinder in the center of the paper, as shown in Figure 7.5. A cylinder in the center of the paper can be conveniently used, such as for aligning several cylinders or combining with other parts.

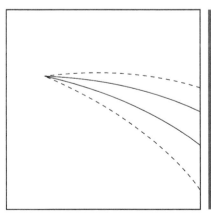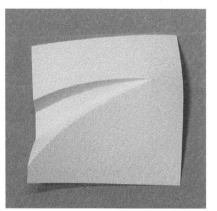

Figure 7.1   Four curves to make a projection (right: backside).

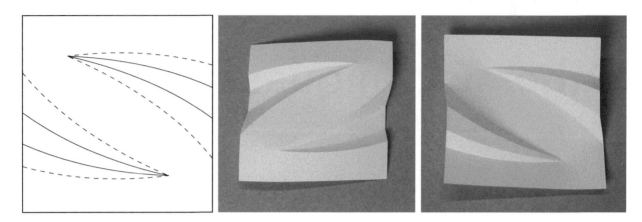

Figure 7.2   Two projections laid out (right: backside).

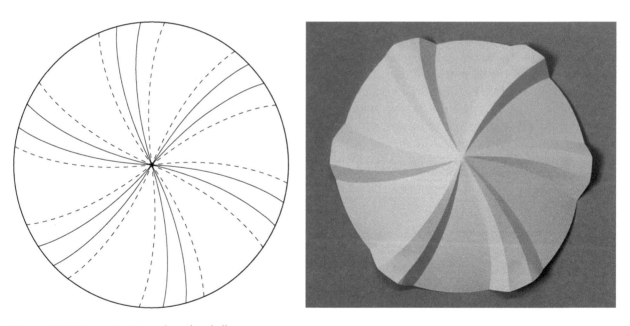

Figure 7.3   Six projections placed radially.

Figure 7.6 is a shape with two cylinders side by side. For this cylinder to be folded easier and more neatly, two vertical fold lines are added inside based on Figure 7.5 crease pattern. This makes the shape sharp after folds.

You may place as many cylinders as you like in different widths and lengths. The top and bottom widths are not necessary the same. By making one end smaller, a cone pops up, as in Figure 7.7.

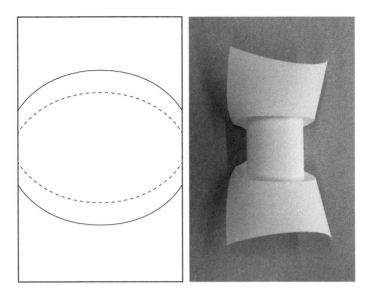

Figure 7.4   Cylinders from aligned curves (crease pattern and photo in Figure 2.18 rotated by 90 degrees).

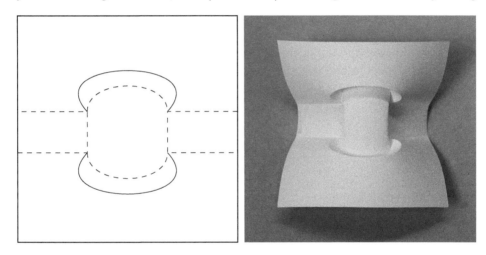

Figure 7.5   Cylinder folded in the center of the paper.

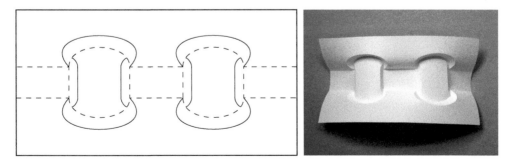

Figure 7.6   Two cylinders aligned side by side.

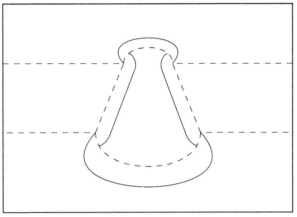
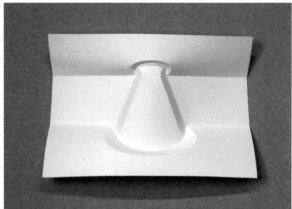

Figure 7.7   Different top and bottom widths make a cone.

7.3 Laying Out Alternatingly

The first half of this book presented examples of aligning the same-shaped curves. In this section, a curve and its reverse are staggered with each other, as in Figure 7.8. To fold

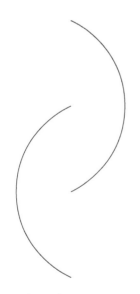

Figure 7.8   Same-shaped curves staggered.

them neatly, valley-fold lines are added, as in Figure 7.9. From a different point of view, this pattern is considered as two aligned pieces of the partial cone presented in Section 5.3. This basic structure can be aligned and connected to create various shapes.

This simple structure can be used as a convenient component combinable into a complex shape. Figure 7.10 has this structure arranged vertically. It makes partial cylinders on both sides. You may use multiple numbers of this structure in vertical alignment to make it taller or reverse it to connect horizontally. Or, you may combine it with other curves to form new shapes.

Figure 7.11 is Figure 7.9 crease pattern aligned to its reverse, with some elaboration so that the fold lines connect smoothly. It's like a mushroom after folds. Let's align this component. In Figure 7.12, the component turned clockwise is connected to its reverse. This forms a three-dimensional, clearly indented shape. Figure 7.12 has four aligned pieces of it. You may align as many components as you like (see Figure 7.13).

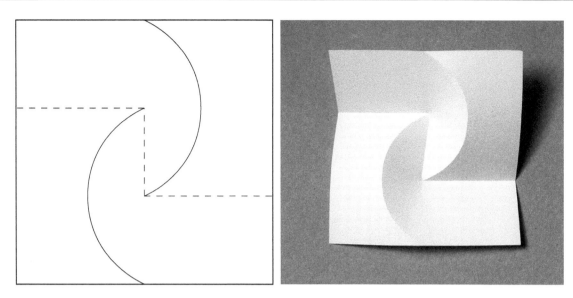

Figure 7.9 Crease pattern added with valley-fold lines and resulting shape (considered as combination of partial cones).

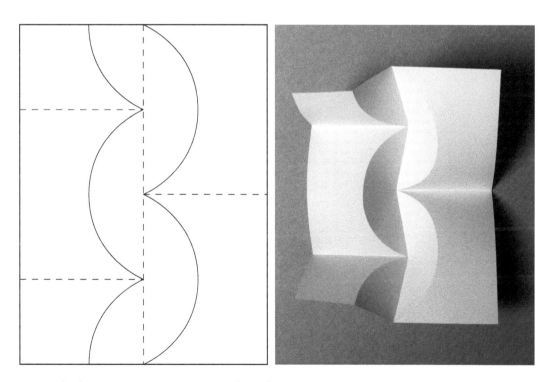

Figure 7.10 Multiple Figure 7.9 crease patterns aligned.

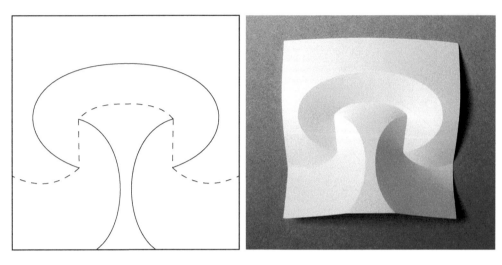

Figure 7.11   New fold line pattern based on Figure 7.9.

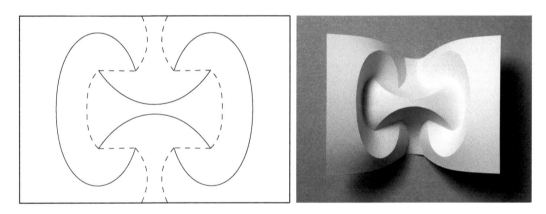

Figure 7.12   Figure 7.11 turned counterclockwise (CCW) by 90 degrees aligned with its reverse.

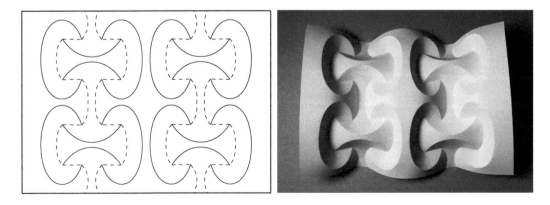

Figure 7.13   Four Figure 7.12 shapes aligned.

7.4 Folding Closed Curves

To neatly fold a closed looping back curve, make a hole in the center. As curves wind sharply after folds, the folded shapes will be greatly distorted. Figure 7.14 has circles in different radiuses from the same center with alternating mountain and valley folds. The innermost circle is cut out. After folds, this symmetrical crease pattern settles in a saddle-like shape. Figure 7.15 is a shape from folding squares in different sizes rotationally shifted and aligned. It has alternating mountain- and valley-fold lines. The indented surface is beautifully emphasized by twisting the whole piece. Aligning and folding various non-circle shapes will create interesting shapes. Give it a try with different shapes other than what is presented here.

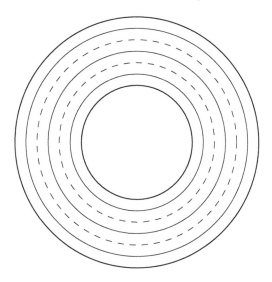

Figure 7.14 Concentric circles hollowed out and folded.

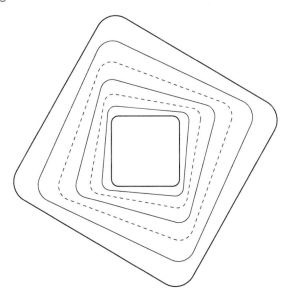

Figure 7.15 Folding squares in different sizes rotationally shifted and aligned.

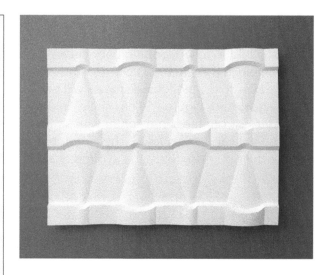

Note

The cylinder folded in Figure 7.5 is utilized for eyes in animal masks by Roy Iwaki. A shape with two staggered parabolas is called a "parabola gadget" in the book titled *Designing with Curved Creases* (Duks Koschitz 2016), though the curves in Figure 7.9 are not a parabola. There is an official record that a modeling similar to Figure 7.14, with alternate concentric mountains and valleys (not hollowed in the center), was created by a student during 1927–1928 in class at Bauhaus provided by German artist Josef Albers (1888–1976). Figure 7.15 was created by reference to the modeling by Aleksandra Chechel.

Artwork 18: Relief of Cones

The same-shaped projections and recesses are aligned at top, middle, and bottom. The middle ones are horizontally shifted against the top and bottom. A faint cone shape emerged from connecting each fold line with a slant-valley-fold line.

Artwork 19: Tiling of Curved Pleat-Folding

This is a pleat-folded pattern consisting of simple curves aligned side by side. The central linear line makes the part of a cone popped up. The finish is a simple and sharp shape.

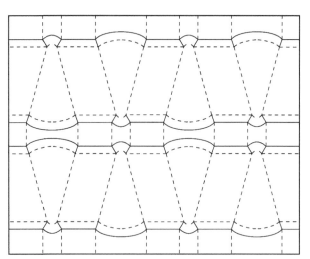

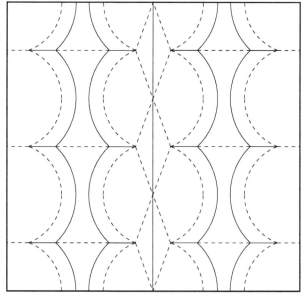

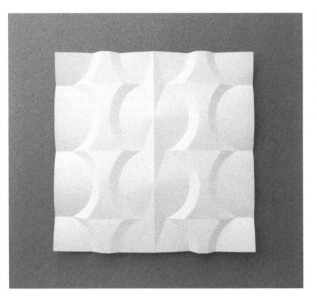

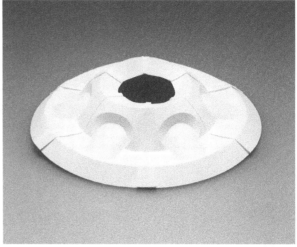

Artwork 20: Wheel

Six pieces of a pattern that folds a cylinder are laid out on a circle. The work has pleats directing toward the center between the cylinder patterns so that it rises conically. The center is cut in a circle for workability. The shape looks like a flying saucer or a burner cap head.

Artwork 21: Japanese Rock Garden 2

This work has a mound in the center with ripples spreading around it. Ripples are folded from alternating mountains and valleys. The tucking technique is applied at several portions. A rock garden-like shape emerges in shading.

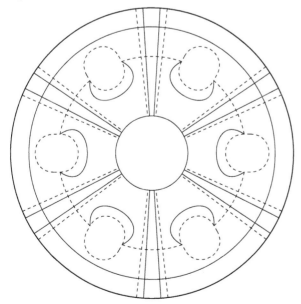

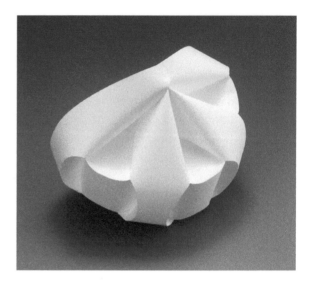

Artwork 22: Variable Transformations of Cylinder Part

What should I call this shape, which emerged during trial and error in laying out curves based on a cylinder component crease pattern? It may be difficult to imagine the finished shape from the crease pattern. Works sometimes are created by chance, even without intended shapes. This is also a joy of creation.

Artwork 23: Rectangles Connected with Tunnels

The cylinder structure in Figure 7.6 is repeatedly placed in grid. The resulting shape has a total of nine soft squares in a three-by-three format. The crease pattern is characterized by valley folds at all intersections of straight lines. These fold lines are also foldable neatly by deftly bending the paper.

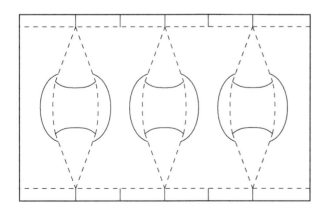

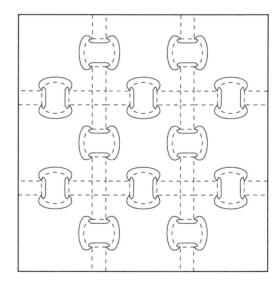

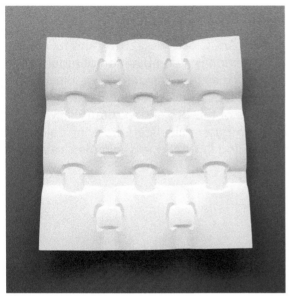

the mountain fold outside and inside of the valley-fold line forms a carved groove.

Artwork 24: Carved Cylinder

The same crease patterns are aligned side by side to make a cylinder as if it is sculpted. Each crease pattern has the staggered arc parts Figure 7.8 connected so that the squiggly valley-fold line makes a circuit. Laying out

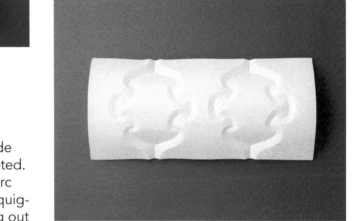

Appendix: Example of Whipped Cream

Figure 7.16 is my work "Whipped Cream" (in *3D Origami Art*, CRC Press, 2016), and its crease pattern when viewed from the inside. Each outside projection is a combination of mountain, valley, valley, and mountain introduced in Section 7.1. This may be considered as a pleat fold based on a mountain-and-valley pair aligned with its reverse, as introduced in Section 2.1. The whole piece becomes rounded by aligning six pieces of this projection rotationally-symmetrically. This shape and its crease pattern were computer-calculated. Its principal is quite basic, such as aligning curves, reversing curves, and aligning in rotationally-symmetrically, as I have presented so far.

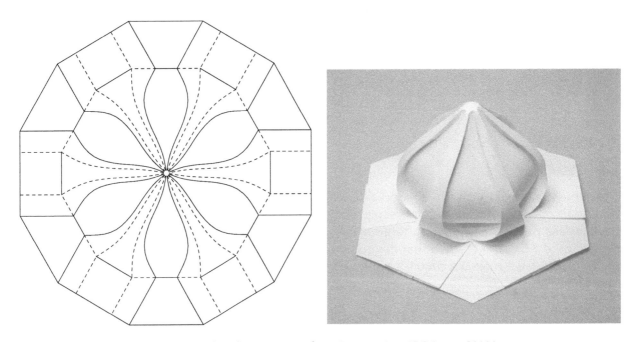

Figure 7.16   Whipped cream-shaped work. (Courtesy of *3D Origami Art*, CRC Press, 2016.)

For Those Who Want More Theory–6

MATHEMATICS OF CURVED FOLDS

Assume that paper is an inelastic material of zero thickness. Then, shapes from bending and folding the paper can be mathematically expressed and calculated based on its geometric constraint. Though shapes after folds with given curves at given angles can be calculated, the inverse problem, i.e., obtaining the crease pattern for the desired shape, is still a challenging task. In most cases, the desired shape cannot be made just from folding paper. Another challenging problem is designing desired shapes that meet the condition of being foldable with a single sheet of paper. On the other hand, human ideas are needed to discover a fold line layout for a beautiful shape. Knowing equations does not directly connect to producing beautifully designed shapes.

This book has demonstrated how to design and create attractive shapes without using equations. Equations may not be necessary for creating shapes other than those presented in this book. Actually, many artists are producing beautiful curved fold modelings without using equations or computer-aided designs. Particularly, for a shape in which the outer circumference is not fixed, the paper settles in a balanced state. This allows you to experiment with this and that to find out interesting shapes.

On the other hand, 3D computer graphics (3DCG) or computer-aided design (CAD) software-based design and analysis needs to replicate paper behavior using a computer. You need to build the shape data in advance when there are restrictions on dimensions or points to fix. This definitely requires equations. Once the possible shape is on the monitor, trial and error is far more efficient than folding a great number of pieces of paper. This section mathematically presents the contents like ruled surfaces, developable surfaces, and curved folds, which are required when computer-handling and calculating the shapes that paper can make. This may not be directly necessary for you, but just take a quick look at the relation between origami and mathematics.

CURVES

CAD and CG use parametric curves in general. For instance, a 2D parametric curve's x and y coordinate values are represented by the equation containing the parameter t. In the equation below, $\mathbf{p}(t)$ represents a semicircle with radius r.

$$\mathbf{p}(t) = (r\cos(t), r\sin(t)), \quad 0 \le t \le \pi$$

The point moving on the curve is determined by the t value. Therefore, the curve is drawn by changing the t value. When discussing the curve properties, including tangent, normal, and curvature, it is very convenient if this parameter value is expressed by the curve length (arc length). The following paragraphs use the arc length parameter s to distinguish from the commonly used parameter representation t. The previous semicircle $\mathbf{p}(t)$ is rewritten as follows using the arc length parameter s.

$$\mathbf{p}(s) = (r\cos(s), r\sin(s)), \quad 0 \le s \le \pi r$$

Generally it is not easy to rewrite an arbitrary parametric curve into an arc length parameter. When handling a curve difficult to represent by arc length parameter, a numerically computed approximate representation is used instead.

Tangent, Principal Normal, and Binormal

The tangent vector of the curve **p**(s) represented by arc length parameter is determined by first-order differentiating **p**. Differentiating the obtained tangent vector gives the acceleration vector, called principal normal. The cross product of the tangent and the principal normal is called binormal. Tangent vector, normal vector, and binomial vector are usually described as **T**, **N**, and **B**, respectively, after their initial letters. The relation between a curve and the vectors are shown in Figure 7.17. The normal vector **N** points to the center of the circle of curvature. Assume that each vector is a unit vector. Then, they are expressed as follows.

$$\mathbf{T} = \mathbf{p}'$$

$$\mathbf{N} = \frac{\mathbf{p}''}{|\mathbf{p}''|}$$

$$\mathbf{B} = \mathbf{T} \times \mathbf{N}$$

p′ is first-order differentiation of **p**(s) and **p**″ second-order differentiation. Differentiation of the curve represented by arc length parameter is a unit length. Therefore, $|\mathbf{p}'| = 1$.

Curvature

Curvature is the degree of curving of a curve at a point on it. This is the reciprocal of the radius of the circle touching the point (circle of curvature). As in Figure 7.18, a large circle touches a gentle curve, and thus the curvature takes a small value. At the point of contact p_0 where a circle

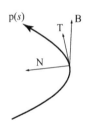

Figure 7.17   Space curve tangent (T), normal (N), and binormal (B).

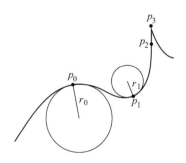

Figure 7.18   Curvature of a curve.

of radius r_0, the curvature is $1/r_0$. A small circle touches a sharp curve, and thus the curvature takes a large value. The curvature is $1/r_1$ at the point of contact p_1 with a circle of radius r_1. The curvature is zero at the straight-section point p_2 and is not defined at unsmooth portions like point p_3. A curve with a constant curvature becomes a circle.

TORSION

Torsion indicates how far a curve is from the plane in three-dimensional space. In the Figure 7.19 curve C_0 is on the plane and its torsion is zero. Other curves are spiral and travel away from the plane. Therefore, their torsions are other than zero. C_2, the fastest traveling one, has the largest torsion of all. The torsion of C_1 and C_2 going in the direction of the right screw takes a positive value, and the torsion of C_3 going in the opposite direction takes a negative value.

Use arc length parameter s and represent the curve $\mathbf{p}(s)$. Then, the torsion $\tau(s)$ is given as follows.

$$\tau(s) = -\mathbf{b}'(s) \cdot \mathbf{n}(s)$$

where $\mathbf{n}(s)$ is the normal vector

$$\mathbf{n}(s) = \frac{\mathbf{p}''(s)}{|\mathbf{p}''(s)|}$$

and $\mathbf{b}(s)$ is the binormal vector

$$\mathbf{b}(s) = \mathbf{p}'(s) \times \mathbf{n}(s).$$

GAUSSIAN CURVATURE

Gaussian curvature is the degree of curving of a curved surface. In general, the Gaussian curvature is the product of two principal curvatures. Figure 7.20 illustrates it intuitively. Set up a normal vector $\mathbf{N_p}$ at the point $\mathbf{p}$ on a curved surface. The bold line is the line intersection between the plane π containing this vector and the original curved surface. The curvature of this line intersection at the point $\mathbf{p}$ changes depending on how the plane π is taken. Rotate the plane π around the normal vector $\mathbf{N_p}$ and observe how the signed curvature changes.

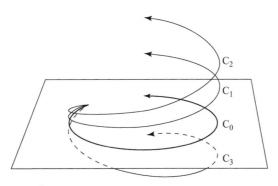

Figure 7.19   Curves with different torsions.

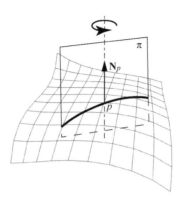

Figure 7.20   Curvature of a curve.

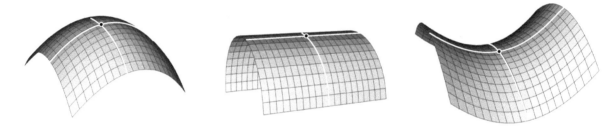

Figure 7.21   Positive and negative Gaussian curvatures (from the left: positive, zero, negative).

The product of the value when the curvature is maximum and the value when the curvature is minimum is the Gaussian curvature. As shown in Figure 7.21, the sign changes by the degree of curving of the curved surface. The Gaussian curvature is positive on a spherical surface and negative on a saddle point. Curved surfaces that can be made from bending the paper are limited to those with the Gaussian curvature of zero.

Ruled Surface

A ruled surface is one of the curved surfaces made by the locus of a straight line traveled in space. More accurately, it is a curved surface made by the locus when the point of a straight line continuously changes by one parameter. To represent how a straight line travels, introduce a smooth parametric curve $\mathbf{p}(s)$ along which the line travels. In other words, setting the parameter s to a value always determines only one line. Represent the line direction by the unit vector $\mathbf{e}(s)$. Then, the position of a point on the line is expressed as follows, which is the ruled surface in parametric representation.

$$X(s,t) = \mathbf{p}(s) + t \cdot \mathbf{e}(s) \tag{7.1}$$

t is the parameter that determines the position on the line. Thus, a ruled surface can be represented as a function of two parameters s and t. Shapes that can be made from bending the paper belong to a ruled surface, because they can be made by aligning lines. On the other hand, not all ruled surfaces can be made with paper.

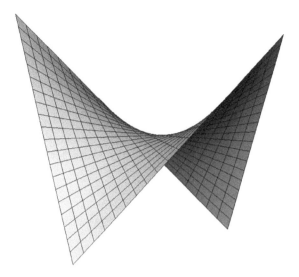

Figure 7.22   Hyperbolic paraboloid.

Figure 7.22 is a hyperbolic paraboloid, one of the ruled surfaces. Suppose we skew lines of p and q. Then, the hyperbolic paraboloid is expressed by $X(s,t) = p(s) + t(q(s) - p(s))$. This cannot be made from bending a sheet of paper.

DEVELOPABLE SURFACE

A developable surface is a curved surface and is isometrically transformable, capable of changing to a plane without compressing or expanding. This is exactly the curved surface made from bending the paper. It is generally a smooth curve with no folds. A developable surface consists of a set of line elements and, thus, is a ruled surface. As shown in Figure 7.23, the normal vectors have the same orientation on the line element contained in the developable surface. The Gaussian curvature is zero at every point on the developable surface. Assume that the curved

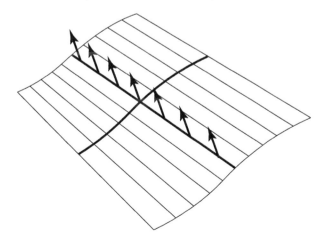

Figure 7.23   Developable surface and line elements. Normal vector direction is constant on a line element.

surface's geodesic line (the shortest curve between two points) is $\mathbf{p}(s)$. Then, the developable surface is expressed by Equation (7.2). This is the same form as the expression (7.1), meaning that developable surfaces are one of ruled surfaces.

$$X(s,\ t) = \mathbf{p}(s) + t \cdot \frac{\mathbf{p}''(s) \times \mathbf{p}'''(s)}{\left|\mathbf{p}''(s)\right|^2} \qquad (7.2)$$

A conical surface has all of its line elements passing through a fixed point. A cylindrical surface has all of its line elements in parallel. A tangent surface is generated by the tangents of a three-dimensional curve. A developable surface is classified into these three types: cone, cylinder, and tangent surface.

CURVED FOLDING

The relationship among the curved-fold line, line element direction, and fold angle is expressed as follows[*].

$$\kappa_{2D}(s) = \kappa(s)\cos\alpha(s) \qquad (7.3)$$

$$\cot\beta_L(s) = \frac{\alpha(s)' - \tau(s)}{\kappa(s)\sin\alpha(s)} \qquad (7.4)$$

$$\cot\beta_R(s) = \frac{-\alpha(s)' - \tau(s)}{\kappa(s)\sin\alpha(s)} \qquad (7.5)$$

$\kappa_{2D}(s)$ is the curvature of a fold line on the plane, $\kappa(s)$ and $\tau(s)$ are the curvature and torsion of the fold line folded (space curve), $\alpha(s)$ is the fold angle, and $\beta_L(s)$ and $\beta_R(s)$ are the angles made by line element and curve on the crease pattern, as illustrated in Figure 7.24.

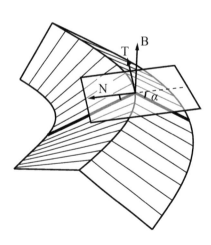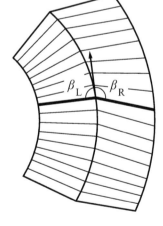

Figure 7.24   Curved folding and line elements.

[*] The equations are from: Tomohiro Tachi. "One-DOF Rigid Foldable Structures from Space Curves." in Proceedings of the IABSE-IASS Symposium 2011, London, UK, September 20–23, 2011.

Equation (7.3) indicates that the 3D curved-fold line after folds has a larger curvature than the curve defined on the plane. In other words, the curve bends larger than the original one after being folded. The larger the bend angle is the more the curvature is. Bending at a larger angle makes a shaper curve, as presented in Figure 1.8.

Equations (7.4) and (7.5) indicate the relationships among the line element, change in fold angle, and torsion. Suppose that the folded curve's torsion is zero, meaning that it gets on the plane, and the fold angle is constant independently from the location ($\alpha'(s) = 0$). Then, $\cot\beta_L(s) = \cot\beta_R(s) = 0$. Therefore, the line elements are perpendicular to the tangent of the curve on the crease pattern. In other words, the line elements are laid out almost perpendicular to the curve if the curve is less twisted and the fold angle remains unchanged at any location (the natural state made by the paper).

Afterword

This book has presented the techniques for curve-folding paper beautifully and neatly, with lots of photographs and sample works. Most of there crease patterns are simple, consisting of several curves. However, the process of preparing these sample works involved much trial and error. It was so exciting to write this book while planning the whole configuration and finding out new ways of folding at the same time.

Earlier in this book, I showed that one single curve can create a variety of shapes. That a braid-like shape emerged from using a paper strip was a discovery to me as well. For the sample works containing tucks, I tried creating shapes by drawing fold lines freehand. The dynamic fold lines that then emerged really surprised me. Among the techniques in the last chapter, the method of aligning arcs in a staggered manner offers a much broader range of applications, contributing to my creativity.

Before writing this book, my origami design style was creating a final shape and its crease pattern through computer calculations, before actually folding the crease pattern. But with the approaches in this book, you never know the results until you really do it. These approaches produced many failures but taught me many things at the same time. I came to recognize again the depth of curved-fold origami. Some shapes are calculated, but there are many more that cannot be mathematically expressed. A combination of theory-based designing and experience-based intuitive designing is necessary for evolving the origami world.

Enjoying origami counts on you trying it yourself. Hands-on experiments give you new discoveries. As you fold the paper, you'll see how your workpiece changes in appearance and shading, continuously or all of a sudden. This magical moment of truth is beyond all words. You can only witness it through your own real experience.

I would be more than happy if this book helped you make the first step into the attractive world of curved-fold origami.

Index

Note: Page numbers in italic refer to figures.

T - #0533 - 071024 - C122 - 254/203/6 - PB - 9780367180256 - Gloss Lamination